Evidence of Energy

An Introduction to Mechanics

Book Two

Jack E. Gartrell, Jr.   Larry E. Schafer

◆

A Project of Horizon Research, Inc.

Materials for middle-grade teachers in physical science

This project was funded by BP America, Inc.

The National Science Teachers Association

Stock Number #PB080X
ISBN # 0-87355-092-7

Printed in the United States of America
by Automated Graphic Systems

EVIDENCE OF ENERGY

Table of Contents

MODULE 1

Pondering Projectile Pathways   **1**

ACTIVITY 1   3

Bent Barrel Ballistics

invites students to investigate the motion of a ball that has been traveling in a circular path inside a piece of tubing. When the ball leaves the tubing, will it continue to curve or go straight?

ACTIVITY 2   7

Tunnel Trajectories

presents students with the task of rolling a tennis ball from one end of a curving tunnel to the other. Can the ball be made to curve, without touching the tunnel walls, after being released at the mouth of the tunnel?

ACTIVITY 3   11

Airmail Bowling Balls

allows students to investigate the motion of a "bowling ball" that is dropped from a moving "airplane."

ACTIVITY 4   17

Pancake Catapult Explorations

presents a series of challenges to help students discover what variables affect the flight of a pancake being fired from a catapult.

MODULE 2

25

It's All in a Day's Work

MODULE 3

49

Energy and Energy Conversions

MODULE 6

Center of Gravity

119

Readings   149

◆Acknowledgements

A project like *Evidence of Energy: An Introduction to Mechanics, Book Two* does not reach publication without the contributions and enthusiastic support of many people. Foremost among these are Iris R. Weiss, Project Director at Horizon Research, Inc., who initiated the program to produce this material, and the authors: Jack Gartrell of Horizon Research and Larry Schafer of Syracuse University.

Karen Johnston provided the outline of concepts to be included, and Ellen Vasu developed the basic design for the modules. Charles Stewart, Norman Anderson, Ruth Sanders, and Charles Beehler provided valuable suggestions for activities and resources to use in developing this material.

The manuscript was reviewed by Fred Goldberg, Robert Gioggia, and Judith Bodnick. They not only verified scientific accuracy, but evaluated the presentation of the concepts involved and the ease with which the activities could be performed. Additionally, John Park and Patti Soule provided valuable feedback after using these materials in an inservice workshop. All these efforts are greatly appreciated.

Sondra Hardis of BP America, Inc., provided many valuable suggestions for disseminating this material, helping to establish links between it and many potential users. Phyllis Marcuccio of the National Science Teachers Association handled the arrangements that made production of this book possible.

Evidence of Energy was produced by NSTA Special Publications: Shirley Watt Ireton, managing editor; Michael Shackelford, assistant editor; Cheryle Shaffer, assistant editor. Michael Shackelford was NSTA editor for *Evidence of Energy*. The design was by Sharri Harris Wolfgang of AURAS Design. Illustrations were created by Sophie Burkheimer and Larry Schafer. The contributions of all these people were essential in the preparation of this book.

Evidence of Energy was a project of Horizon Research, Inc. Project Director was Iris R. Weiss; Project Co-director was Ann C. Howe.

Special thanks go to BP America, Inc., for providing the funds to make this project possible.

OVERVIEW

Evidence of Energy
An Introduction to Mechanics

Mechanics is the branch of physical science that deals with energy and forces and their effects on bodies, providing us with a way to describe and predict the behavior of bodies in motion. Much of our knowledge of mechanics is based on the work of Sir Isaac Newton. In 1687, Newton published a brilliant synthesis of the principles of mechanics—*Philosophiae Naturalis Principia Mathematica* (commonly known as *Principia*). Newton's insights were so powerful and complete that they remained virtually unchallenged and unmodified for two centuries. This is the second in a two-volume set of modules designed to illustrate and explain Newton's work in a way that will be useful to teachers who work with students in the middle grades. Book One, *Methods of Motion*, concentrates on Newton's laws of motion. While this volume builds upon some of the fundamental concepts presented in the first, the activities presented in *Evidence of Energy* are designed to be useful without any previous experience with mechanics. A brief review of the essential ideas from *Methods of Motion* is found in this book's first reading, and the important vocabulary terms from the first book are also included in this glossary.

Introducing mechanics in the classroom is often difficult. It sometimes seems that no matter what aspect of the topic the teacher presents, the students need advance knowledge of other information before dealing with the concept at hand. Beginning in the middle seems inescapable.

To further complicate matters, everyday observations of moving objects seem to contradict the unifying principles of mechanics. Newton's first law of motion states that an object in motion will continue in motion at a constant velocity in a straight line unless acted upon by an unequal force. But any sixth grader can tell you that when you throw a baseball, it soon curves to Earth and stops moving all by itself. No reasonable person believes that the ball really would keep moving in a straight line at a constant speed, no matter what the laws of motion may predict. All of our experience suggests otherwise.

◆Organization of this book

There is no perfect solution to the problem of how to begin studying motion. Neither is there a single best method to correct people's misconceptions about the behavior of objects that seem to disobey the laws of motion. The approach used in this series of modules is to illustrate selected concepts of mechanics with hands-on activities and audiovisual materials. Several segments of the *Eureka!* video series, produced by TVOntario, are used for concept presentations.

Following the modules is a collection of readings. This section gives detailed explanations of the concepts presented in the activities. The readings also can be reproduced as student handouts to supplement textbook materials or augment student discussion at the completion of the activities. Permission to reproduce both the readings and the activities for classroom use is given by NSTA.

A guide for teachers and workshop leaders is provided for use in planning instruction of the activities. This section lists equipment and materials required to perform each module's activities. Ordering information for the recommended audiovisual materials is also included in the guide for teachers and workshop leaders. Finally, a glossary provides definitions of many of the terms used in these modules. This glossary also serves as a master list of vocabulary introduced and defined in the activities.

◆Getting ready for classroom instruction

Most of the activities in this book are intended to be hands-on experiences in which the students can actively participate. Some activities function better or *must be* performed as demonstrations, but even those intended as hands-on activities can easily be adapted for classroom demonstrations.

Each module begins with a discussion of the rationale, objectives, and overview of the activities within it. Each activity begins with a reproducible student worksheet, which also contains the concept objective and any new vocabulary words. The activities are designed to take between one and two class periods (40–60 minutes) to complete, unless otherwise indicated.

Each activity worksheet is followed by a commentary for teachers called "Guide to Activity. . . ." These sections explain the expected results for the activity and provide additional background information on the content of the lesson. You also will find suggestions for time management, teacher preparation, ways to help students obtain reproducible results, ways of troubleshooting equipment, and hints about problems that may be encountered while performing the activity. Safety notes, sample data, and answers to the questions on the activity worksheet are also provided. In addition, these commentaries contain suggestions for further study, outlining other experiments that can be performed using the same apparatus.

The equipment required for the activities consists mainly of inexpensive toys and other low-cost, readily available materials. Good results can be obtained for many activities by using alternative procedures or by using substitutes for materials listed on the worksheet. If you do not have ready access to the materials listed, the "Guide to Activity. . ." section offers possible substitutions and procedure modifications.

You will note that these modules use metric units wherever possible. These are the units routinely used in science. Length is typically measured in meters (m). One meter is equivalent to 39.37 inches, or slightly more than a yard. Shorter lengths are measured in centimeters (cm) or millimeters (mm). One meter is equal to 100 cm or 1000 mm. Large distances are measured in kilometers (km). One kilometer is equal to 1000 m, or about 0.7 of a mile.

Volume is typically measured in liters (L). One liter is slightly more than a quart. A gallon is equal to 3.8 L. Small volumes are usually measured in milliliters (ml). One L is equal to 1000 ml. One teaspoon of a liquid is equal to about 5 ml.

Time is typically measured in seconds (s). Other units familiar from the English system are occasionally used, but the second is the most common.

The metric units for mass, the kilogram (kg), and force, the newton

(N), are discussed in the modules and readings. One thousand grams (g) is equal to 1 kg.

In many cases the English system equivalent is provided to help you and your students make the conversions. We have also included a metric conversion chart at the end of these modules.

◆Getting ready for workshops

If you are using this book as a workshop leader or teacher participant, you may wish to perform some of the activities as classroom demonstrations. As you work through these activities with your colleagues, you will have the opportunity to discuss new insights, explore alternative procedures, and make note of any problems you encounter. These experiences will add to your confidence when you direct your students in similar activities later.

Each module is designed to take one inservice workshop time period. The general topic of each module is described in its introduction. Each introduction lists instructional objectives for the workshop, gives the titles of the activities in the module, and indicates the readings that should be studied after the module's activities are performed.

MODULE 1

Pondering Projectile Pathways

◆Introduction

• Why do basketball shots arch toward the basket?

• Would a ball thrown from a rotating merry-go-round go straight?

• Why do curveballs and fastballs follow different paths through the air?

Many people's ideas about motion are consistent with what is seen in cartoons, video games, and comedy films. In a typical example, Wile E. Coyote zooms straight off the edge of a cliff, hangs motionless for a moment, and then falls straight down to produce a muffled "poof" at the bottom of a deep canyon. While most people realize that objects do not behave as Wile E. Coyote does, their ideas about motion are often closer to "cartoon" motion than to actual motion.

This module takes into account current research about people's conceptions of projectile motion. Researchers have found that a person may appear to understand how objects move, flawlessly reciting Newton's three laws and even using equations of motion to solve rather complex problems, but still reveal an underlying set of misconceptions when asked to perform certain tasks.

The activities in this module are designed to "tease out" any misconceptions that students might have and to provide experiences and arguments that can lead to a more accurate understanding of projectile motion and the concepts used to explain it.

◆Instructional Objectives

After completing the activities and readings for Module 1, students should be able to

• accurately predict the path of an object that is:

(1) pushed along a curving path, then released [Activities 1 and 2]

(2) dropped from a moving platform [Activity 3]

• construct and use a simple catapult, and predict the path of a projectile fired from it [Activity 4]

◆Preparation

Study the following readings for Module 1:

Reading 1: Reviewing Newtonian Mechanics

Reading 2: Projectile Motion

Reading 3: Dissecting Projectile Motion

Reading 4: Psyching Out the Psycho Cyclist

◆Activities

This module includes the following activities:

Activity 1: Bent Barrel Ballistics

Activity 2: Tunnel Trajectories

Activity 3: Airmail Bowling Balls

Activity 4: Pancake Catapult Explorations

ACTIVITY 1 WORKSHEET

Bent Barrel Ballistics

◆Background

In an old Bing Crosby and Bob Hope movie, *The Road to Bali*, Bing stands in the doorway of a hut and fires a gun, hoping to escape. Unfortunately for Bob and Bing, the barrel of the gun has been bent into a semicircle by a gorilla, so the bullet traps the famous twosome by continuously circling the hut. Would you expect this to happen? This activity will give you a chance to see for yourself.

◆Objective

To discover if objects restricted to traveling in a curved path continue to follow a curved path when released

◆Procedure

1. Coil the tubing into a spiral with a 20-cm diameter and tape it down as shown in the diagram. Check the following:

• Is the exit end of the coil taped so that it follows the curve of the coil?

• Is the loading end of the coil taped so that it can be lifted about 10 cm above the table?

• Is the exit end placed so that the ball can roll at least 30 cm beyond it?

• Is the coil taped so that it will not move with repeated shootings?

• Is the bullet small enough to roll easily inside the tubing?

Materials

Each group will need

• 115-cm length of clear plastic tubing (1" OD, 3/4" ID)

• a "bullet"— a marble or ball bearing that can roll freely inside the tubing

• masking tape

• 4 or 5 index card targets about 3 cm wide (one for each person in the group)

• a meter stick

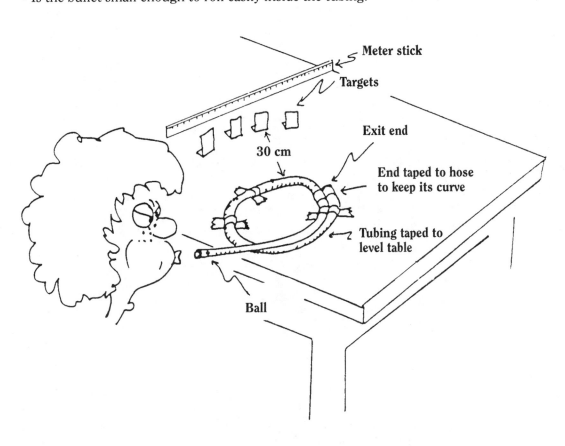

2. Write your name on one of the index card targets. Predict the path of the bullet and try to place your target on the table so that it will be hit by the ball after the ball rolls out of the tube. The target must be at least 30 cm from the exit end of the tube.

3. Raise the loading end of the tube about 10 cm above the table and drop the bullet in the opening. Observe the path of the ball before and after leaving the tube. Whose target was hit?

_____

4. Remove the targets. Place a meter stick about 30 cm in front of the exit end of the tube so that it stops the balls after they roll out of the tube. Shoot the bullet at least five more times and keep track of where the bullet hits the meter stick.

5. After observing where the ball strikes the meter stick in relation to the exit end of the tube, describe the motion of the bullet after it leaves the tube.

_____

_____

6. Will objects continue to follow a curved path after they are free to move without constraint? State your own hypothesis.

_____

_____

GUIDE TO ACTIVITY 1

Bent Barrel Ballistics

◆What is happening?

Even though the marble travels in a circular path inside the tubing, it rolls in a straight line as soon as it leaves the tube. Students may predict that the ball will continue moving along a curved path after it leaves the tube. Identifying the forces acting on the marble shows why this cannot happen.

The marble travels in a circle inside the tube because the walls of the tube exert an unbalanced force on the marble toward the center of the circle. (An unbalanced or unequal force acting on an object is a force that is not canceled out or balanced by any other force or combination of forces acting on the object.) Newton's second law (Force = mass x acceleration, or $F = ma$) tells us that when there is an unbalanced force acting on an object (F), the object will have an acceleration (a), inversely proportional to its mass (m), in the same direction as that of the force. So the object tends to move in the direction of the force. After the marble leaves the tube, the tube does not exert any force on it, and the marble travels in a straight line.

The marble curves *only* when there is an unequal force acting on it. In other words, the marble behaves exactly as predicted by Newton's first law: An object in motion tends to stay in motion in a *straight line* and at a constant speed *unless acted upon by an unequal force.* (See Reading 1 for a discussion of Newton's laws and other fundamentals of mechanics.)

◆Time management

One class period (40–60 minutes) should be enough time to complete the activity and discuss the results.

◆Preparation

Any ball that rolls freely inside the tube can be substituted for the marble or ball bearing. Balls formed out of clay also perform well. Do not use any form of "ammunition" that fits in the tube tightly—the ball may exit the tube at too high a speed or get stuck.

If clear plastic tubing is not available, use short lengths of garden hose.

◆Suggestions for further study

If a flat, smooth table is used in the above activity, the ball leaves the tube and generally travels in a straight line. Explore different things that might be done to get the ball to move along a curved path after it leaves the tube. (Hint: Explore the third dimension.) What interacts with the ball to cause it to move in a curved path outside the tube?

Place the exit end of the tube at the edge of a table. Tape the coil so that it cannot move. Make a penny-sized dot in the middle of a sheet of paper. Lay the paper on the floor so the dot is where you predict the ball will land. Test out your prediction. Without changing the direction of the exit end of the tube, try to get the ball to land in front of that dot and place a mark where the ball lands. Then get the ball to land behind that dot and mark that landing position. Do the three dots describe a definite curve or more of a straight line?

Obviously the ball curves downward toward the floor as it falls. What causes that curving motion? How might it be possible to say that the ball is both traveling in a curve and in a straight line at the same time?

◆**Answers**

5. The bullet travels in a straight line after it exits the tube, even though it has been traveling in a circular path while inside the tube.

6. Most observers will say that an object moves in a straight line unless it is forced to move in a curve (by something such as the tubing). This hypothesis restates Newton's first law: An object in motion tends to stay in motion in a *straight line* and at a constant speed *unless acted upon by an unequal force.*

Some students may still predict a curved path for the projectile after it leaves the tube. Remind them of their observations: As soon as the ball leaves the tube, it goes in a straight line.

ACTIVITY 2 WORKSHEET

Tunnel Trajectories

◆Background

How do pitchers throw curve balls? Why do basketball shots arch? To make an object follow a curved path, a sideways **force** has to be applied to it that alters its forward motion. Can you apply a sideways force to a tennis ball so that it rolls inside a curved tunnel without touching the walls?

◆Objective

To discover whether and how an object can be made to follow a curved path

◆Procedure

1. Draw two arcs on the paper as shown—one with a radius of 30 cm and one with a radius of 55 cm. Bend each of the index cards so as to form a base by which the cards can be taped to the paper. Tape the index cards along the arcs as shown in the drawing. The cards represent the walls of the tunnel and make it easier to see if the ball stays within the tunnel.

Materials

Each group will need

• a sheet of paper measuring at least 60 cm x 60 cm

• a felt-tip pen or other marker

• eight 3 x 5 inch index cards

• masking tape

• a tennis ball

• a meter stick

Vocabulary

• **Force:** A push or pull in a particular direction that can be applied to an object. We can more technically define a force as something that has the capacity to change the motion of an object. Both the *magnitude* (strength) and the *direction* must be stated when defining a force. Vectors are often used to represent forces.

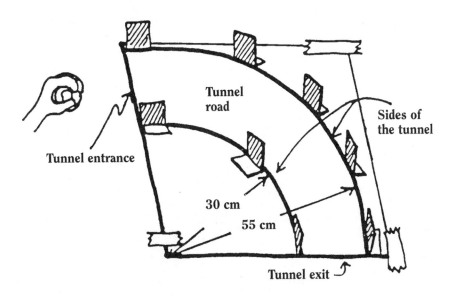

Tunnel road

Tunnel entrance

Sides of the tunnel

30 cm

55 cm

Tunnel exit

2. The challenge is to make the ball roll into the entrance and out of the exit of the tunnel while following these rules:

• The ball must stay within the walls of the tunnel.

• You must release the ball before it enters the tunnel.

• The ball must roll—no bouncing allowed.

• The ball must not spin—no English allowed.

3. Move the ball along a curved path on a level surface, then let it go. What path does the ball take after being released?

4. Try different methods of rolling the ball the length of the tunnel and describe the path of the ball after it is released. Remember: *No bouncing or spinning is allowed.*

5. Based on your observations, does the path of the ball after it is released depend on how the ball is moved before it is released?

_____

6. Suppose you observe someone move the ball along a curved path and then release it. Explain the ball's motion if it:

(a) continues a curved motion after release.

_____

_____

(b) rolls in a straight line after release.

_____

_____

GUIDE TO ACTIVITY 2

Tunnel Trajectories

◆What is happening?

If the rules are followed, the challenge can only be met by rolling the tennis ball on a straight line from the outside edge of the tunnel entrance to the outside edge of the tunnel exit. Attempts to get the ball to roll in a circular path by moving it in a circle before releasing it will not be successful.

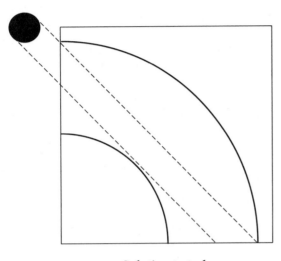

Solution to task

The purpose of the activity is to give students further hands-on experience with Newton's first law: An object in motion tends to stay in motion *in a straight line* and at a constant speed unless acted upon by an unequal force. The activity is designed to emphasize the importance of considering unequal forces when applying the law. The hand can provide an unequal sideways force to move the ball in a curved path before release. However, the ball travels in a straight line after release since the sideways force is only exerted while the hand touches the ball. An unequal force can only alter the straight-line, constant-speed motion of an object while it is being applied—not after.

It is possible for persons with a very deft touch to curve the ball through the tunnel by spinning it like a top in addition to giving it a push into the entrance of the tunnel. The spin on the ball (or English, as pool players call it) exerts a sideways force which continues to operate after the hand is no longer in contact with the ball. Thus, an unequal force continues to be applied after release, causing the ball to curve in the direction of the spin. Applying English, however, is prohibited by the rules. The spin and the force it produces complicate analysis of the ball's motion and obscure the illustration of the first law.

◆Time management

Half of one class period (20–30 minutes) should be enough time to complete the activity and discuss the results.

◆Preparation

Be sure that the table surface you use for this activity is level. If the table slopes very much, the results can be difficult to analyze. It is also important to enforce the rules against spinning and bouncing for at least the students' initial attempts in order to give as clear an illustration of the first law as possible.

◆Suggestions for further study

After doing this activity, students may wish to explore how spinning an object can produce a force that alters what would otherwise be straight-line motion, particularly in the context of baseball. The results of a systematic investigation of whether and/or how a curveball curves are reported in "Pitching Rainbows: The Untold Physics of the Curve Ball," the cover story of the October issue of *Science 82*. The article is fun for the baseball buff and contains many striking pictures of balls stopped photographically by a strobe light.

◆Answers

3. The ball travels in a straight line as soon as it is released.

4. The ball travels in a straight line as soon as it is released.

5. No. The ball always travels in a straight line as soon as it is released as long as the rules are followed.

6. Both cases are explained by Newton's first law. In condition (a), the curved motion means that there is an unequal force operating on the ball; otherwise the ball would travel in a straight line at a constant speed. If the table surface is unlevel, gravity will produce an unbalanced force. Other possible sources of an unequal force include bouncing and spinning the ball. Anything producing an unequal force would explain condition (a).

In condition (b), the straight-line motion means that there are no (or at least only small, negligible) unequal forces present. If the rules of the activity are followed, no unequal forces will be present to alter the straight-line motion Newton's first law predicts.

ACTIVITY 3 WORKSHEET

Airmail Bowling Balls

◆Background

In some of the expansive, sparsely populated states, mail is delivered by airplane. The plane flies low over a home and the letter carrier tosses a bundled package of mail to the ground. Suppose a rancher sends for a bowling ball. The ball comes in the mail and is delivered by the usual method (bombs away).

Since the ball is spherical and massive, it is affected by **air resistance** even less than a heavy mailbag would be. In order to make sure the ball lands near the mailbox, the ace letter carrier must think carefully about when the ball should be dropped. Casting aside thoughts of "rain, snow, sleet, hail, and the dark of night," the letter carrier reasons: "Since air won't have much effect on the ball, the only force on the ball is that of **gravity**. Since the force of gravity on the ball acts straight down, the ball must fall straight down. So, as I fly by I will drop the ball when I am directly over the mailbox."

Is the reasoning of the flying letter carrier correct? Would the bowling ball land near the mailbox?

◆Objective

To simulate the fall of the letter carrier's bowling ball in order to investigate **projectile** motion

◆Procedure

1. Assign each member of the group a role. You need a string holder, a flier, and one or two observers.

2. Tie a loop in each end of the string. One loop should fit loosely over the end of the popsicle stick, and the other over a finger of the string holder.

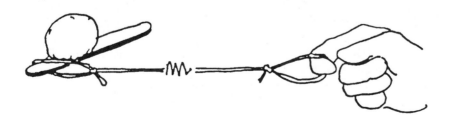

Materials

Each group will need
- a ball of clay about the size of a golf ball (the bowling ball)
- a popsicle stick (the airplane wing)
- a plastic cereal bowl or a half sheet of paper (the mailbox)
- a piece of string about 4 m long (the trip line)
- 10 pieces of masking tape, each 1 cm long
- a meter stick

Vocabulary

- **Air resistance:** A force exerted on a moving object opposite to its direction of motion due to the friction between the object and air. Air resistance is also called *drag* or *air friction*.

- **Gravity:** The force due to gravity is the force of attraction that exists between all objects in the universe. The amount of gravitational force between two bodies (such as the Earth and a rock thrown up into the air) depends on the mass of both objects and the distance between them. Earth's gravitational force is just one example of the general phenomenon of gravity. On Earth, the force due to gravity is the force that causes objects (such as an airborne rock) to accelerate toward the Earth.

- **Projectile:** An object cast or thrown by an external force. Its motion continues because of its own inertia.

3. Clear a "flight path" about 1.5 m wide and about 6 m long. Have the string holder stand at the beginning of the flight path. Place the mailbox toward the end of the flight path so that the popsicle stick is over it when the string is pulled taut.

4. When everything is set, the flier should stand close in front of the holder and check the following:

• Is the string looped loosely over the end of the stick?

• Is the clay ball slightly flattened on one side?

• Is the clay ball resting, *not stuck*, near the end of the stick?

• Is the stick (with ball) parallel to the floor?

After completing the checklist, the flier should begin walking very rapidly toward the mailbox.

5. When the wing passes directly over the mailbox, the string will tighten and jerk the wing from under the ball. *The flier should continue flying at the same speed*, just as the airplane would continue after the delivery of the bowling ball.

Since the ball begins its fall at the moment the string tightens, the ball will be dropped directly over the mailbox. Where will the ball land? Where would the ball land if you shortened the string so that the ball dropped before the wing was over the mailbox?

6. Carry out the mail run five times. Make sure the flier walks as close as possible to the *same speed* each time. For each run, have the observers mark with tape:

• where the ball lands

• where the flier is when the ball lands (remember, the flier should continue walking at the same speed after the string is pulled taut).

7. Based on what you have observed, try changing the length of the string so that the bowling ball will land on (or near) the mailbox. Measure the amount by which you lengthen or shorten the string.

8. Make a drawing showing the path the ball takes while moving through the air to land on the ground target.

9. If an object is moving horizontally and is dropped, will it
• fall straight down with no more horizontal motion?
or,
• fall but continue its horizontal motion?

GUIDE TO ACTIVITY 3

Airmail Bowling Balls

◆What is happening?

When the bowling ball is released, it is traveling at exactly the same speed as the plane. No force is pushing the ball forward; no force is necessary to keep the ball moving. It continues moving in the same direction and at the same constant speed (the speed at the moment of release) as the airplane because of inertia (Newton's first law). Since the friction between the ball and the air is (assumed to be) negligible, there is no horizontal force acting on the ball to change its forward motion, and the bowling ball continues moving forward at the same speed until it strikes the ground. It does begin to accelerate downward because there is a vertical force acting on it: The force of gravity accelerates the ball toward the Earth at a rate of about 10 m/s².

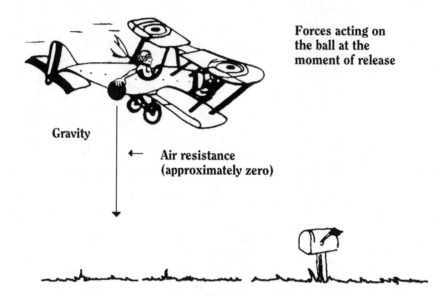

Forces acting on the ball at the moment of release

Gravity

← **Air resistance (approximately zero)**

The observations made during the activity show that when the clay ball is dropped from the moving popsicle stick it does not fall straight down. Instead, it continues its forward motion while it falls due to the influence of gravity. (The force of gravity was acting on the ball even before it was dropped. However, before the ball's release, the force of gravity acting on the ball was balanced by an upward force on the ball exerted by the stick. After the release, the force of gravity was not balanced by any other force and therefore caused the ball to accelerate downward.) If the ball is dropped from a moving airplane directly over the ground target, it lands beyond the target. This observation provides evidence of its continued forward motion.

When the ball is being carried, it has the same forward speed as the flier. When the ball is released, it maintains that same forward motion as it falls. If both the ball and the flier travel with the same forward motion, then they should stay together as the ball falls. Therefore, the ball lands at the feet of the flier. Correspondingly, since the force of friction being exerted by the air striking the bowling ball is very small in our airmail special delivery case, the bowling ball will travel with the same forward motion as the plane after it is dropped. The pilot will have to drop the ball *before* he is over the mailbox; since the ball will continue to move forward as it falls downward, the pilot must consider the forward distance the ball will travel in the time it takes to reach the ground in order to time his drop correctly. The pilot will always be directly over the ball; so, the ball will hit the mailbox just as he flies over it.

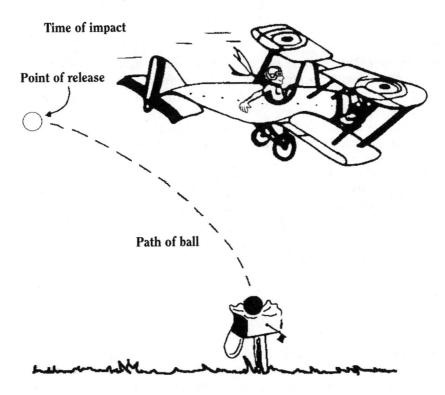

Once an object is set in motion in a particular direction, it will continue in a straight line and at a constant speed unless acted upon by an unequal force (Newton again). Once the ball acquires a constant rate of forward motion, that forward motion *no longer depends on the airplane*. If it did depend on the airplane, the ball's forward motion would stop as soon as it left the plane, and the ball would fall straight down. We know from our observations that this does not happen.

◆Time management

One class period (40–60 minutes) should be enough time to complete the activity and discuss the results.

◆Preparation

Any type of clay (either plasticine, Play-Doh™, or real clay) will work for this activity. Whichever type of clay you use, be sure *not* to stick it down on the popsicle stick. It should simply rest on the stick in such a way that the stick can be jerked smoothly out from under it. You should also make sure that the loop around the popsicle stick is loose enough to slip off the stick when the string is pulled taut so that the flier can continue his or her forward motion. If this presents a problem, have the flier release the stick when he or she feels the string pull and continue walking (*at the same speed*).

◆Suggestions for further study

Have someone sit in a chair that can be smoothly rolled across the floor. Then have the rider lift his or her feet off the floor and use them to hold an open container (an empty coffee can works best). Push the chair and rider at a constant speed. While moving, have the rider try to drop a small ball or marble into the container from a height of about two feet above the container. Should the rider hold the ball in front of, behind, or directly over the container just before the release? Recall that the ball is not in contact with the chair as it falls the two-foot distance from the hand to the can. Where do you hold the ball to get it to drop into the container?

Assume you are riding in a commercial airplane which is traveling at a constant speed of 200 miles per hour. You get a little crazy and decide to stand on your arm rest and jump to the floor. It will take you about one-half a second to fall to the floor. During that time you will not be in contact with the plane and hence your motion cannot be sustained by the airplane. Which of the following should happen?

•You will drop directly to the floor with no backward movement.

•As you fall, the plane will move forward by about 125 feet, but without you since you are no longer in contact with the plane. In other words, you will move 125 feet backwards along the aisle as you fall. (Pick a long plane for this experiment.)

◆Answers

5–9. The forward distance that the ball travels depends on the speed of the person carrying it. Walking at a consistent, fast pace ensures repeatable results. If the person flying continues walking forward at a steady pace after the string tightens and launches the ball, the ball should land at the pilot's feet. The ball always lands somewhere beyond the point of release. If it lands beyond the target, shorten the string.

5. The ball will land beyond the target, at the feet of the flier. If the string is shortened, the ball will land closer to the target. With the proper string length, the ball can begin its fall early enough to land directly on the target.

6. The observations will show that the flier and the ball stay together, since the ball maintains the same forward motion as the flier.

7. The change in the string length will vary depending upon the walking rate of the flier.

8. The path the ball takes is best represented by a smooth curve (see the illustration in the "What is happening?" section).

9. If an object is moving horizontally and is dropped, it will fall but continue its horizontal motion.

ACTIVITY 4 WORKSHEET

Pancake Catapults

◆Background

When you fire a slingshot or pitch a baseball, what determines the path your projectile will take? Understanding the path a projectile takes requires an understanding not only of the forces acting on the object, the

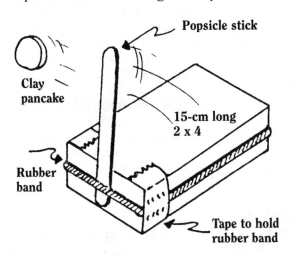

initial motion of the object when it is fired, and the object's **inertia**, but also an understanding of how these factors affect one another. If an object is traveling horizontally before the force of gravity starts pulling it downward, we can easily visualize a flight path that curves downward as it continues forward. But what if a projectile starts out by moving at an

upward angle away from the Earth? How do gravity and air resistance affect its motion? What would its flight path look like?

◆Objective

To build a catapult that fires clay pancakes in order to examine how a projectile's firing angle affects its flight path

◆Procedure

Part I   Preparing your catapult

1. Place your rubber band around the block of wood. The top of the rubber band should be exactly half way up from the bottom of the block. Check to be sure it is even and flat all the way around. Tape the rubber band to the block to keep it in place.

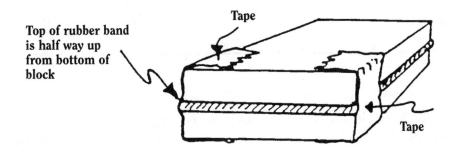

2. Mark your popsicle stick with two lines: a rubber band line 2 cm from one end of the stick, and a pancake line 0.5 cm from the other end. The top of the rubber band will rest on the rubber band line, and the edge of the pancake will rest on the pancake line.

Materials

Each group will need:

• goggles or safety glasses with side shields for each member of the group

• a piece of 2" x 4" lumber about 15 cm long

• a popsicle stick and rubber band (use a medium-wide to wide rubber band at least 4 mm wide and 8.5 cm long)

• masking tape

• clay

• a meter stick

• a book to serve as a base for the catapult

• an index card

• a protractor

Vocabulary

• **Inertia:** A measure of an object's resistance to change in motion. Inertia is another way of describing an object's mass. The more mass that an object possesses, the more force that is required to set it in motion *or* to stop it from moving. Inertia is a property possessed by all matter that can be thought of as laziness or "difficult-to-moveness."

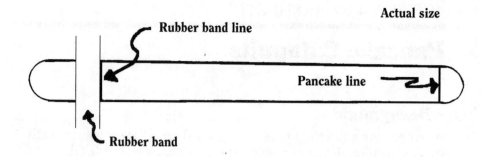

Actual size

Rubber band line

Pancake line

Rubber band

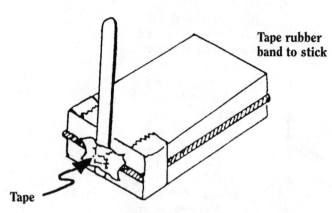

Tape rubber band to stick

Tape

Place the stick behind the rubber band. Tape the rubber band to the stick as shown. This will prevent the stick from accidentally flying away from the catapult.

3. Your pancake should be at least as large as a quarter and about twice as thick. See the suggested actual size below. Make sure your pancake is flat.

4. Use a pencil to make a mark at the edge of the pancake. This mark will be used to place the pancake on the stick consistently for all trials. The mark will represent the back of the pancake and should always be aligned with the pancake line on the popsicle stick.

5. Place—*do not press*—the pancake on the stick. If you press, the pancake will adhere to the stick and not fire properly. There usually is enough "stickiness" to keep the pancake from sliding on the stick even when the catapult is angled upward. If the pancake doesn't stick without pressing, roll a piece of clay along the stick, and try again.

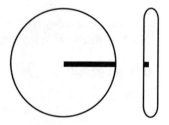

Actual size of pancake

Mark to designate back of pancake

> **Everyone should wear safety glasses or goggles while performing this activity! Don't shoot at anyone!**

> **Tape the rubber band to the stick to prevent the stick from flying away from the catapult. The top of the rubber band must not be lower than half way down the block.**

> **Don't shoot anything except the clay pancakes! Don't shoot pancakes smaller than quarters!**

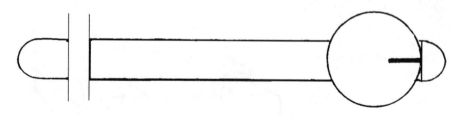

6. The following roles should be assigned in each group:

• **Leader/loader:** Makes sure that everyone is involved. Sees that others are performing their roles. Starts and stops group activity. Keeps the group on schedule. Loads the catapult by placing the pancake on the stick.

• **Shooter/safety officer:** Brings back the arm of the catapult and releases the arm to fire the catapult. Makes sure no one is in a position to be hit by the pancake. Checks to see that the arm is taped to the rubber band.

• **Checker/data manager:** Makes sure that the firing is performed the same way each trial. Makes sure that firing angles are held constant. Records the necessary data.

• **Marker/returner:** Marks the landing points of the pancake. Returns the pancake to the leader/loader. Can stop the shooting at any time. Works with the checker/data manager.

Part II   Testing the catapult

7. Set the catapult at a 45° angle by resting it over the spine of a book. Stick a piece of masking tape on the book and up over the back of the catapult (see below). The tape should hold the catapult at the determined angle. Make sure the base of the catapult is pulled back against the tape.

To prevent the spine of the book from twisting, tape the book shut both ways at one corner.

The tape can be adjusted to hold the catapult at any predetermined angle. Use a protractor to measure the angle between the book and the bottom of the block. Note: *The angle of the catapult is not necessarily the angle at which the pancake takes off from the stick.*

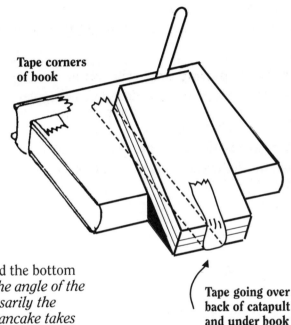

Tape corners of book

Tape going over back of catapult and under book

8. *Put on your safety goggles*. Observing *all the safety rules*, fire the catapult. When you are trying to get consistent results, always bring the stick all the way down to the block. Hold the stick at the very end. Don't jerk or press down when you release the stick. Mark the spot where the pancake lands with a small piece of masking tape.

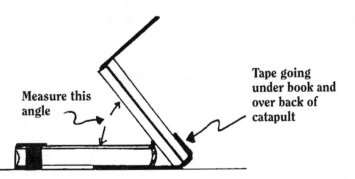

Measure this angle

Tape going under book and over back of catapult

9. Measure the distance (in centimeters) from the front of the base of the catapult to the tape marker. Shoot the catapult a total of ten times without changing the angle. Mark each landing point and record the distance for each shot in the data table.

Data table

Distance pancake traveled

Trial #

1 _____   2 _____   3 _____   4 _____   5 _____

6 _____   7 _____   8 _____   9 _____   10 _____

Total distance _____

10. Find the average distance for the ten shots.

Average distance = Total distance for all 10 trials/10 = _____ cm.

Part III   Exploring projectile paths

11. Try launching your catapult at different angles. What is the greatest *horizontal distance* (measured from the front of the catapult) you can send the pancake? _____ cm

What angle resulted in the greatest distance? _____ °

Sketch this set-up, showing the angle, distance traveled, and the pancake's flight path.

12. Continue trying different angle settings. What is the greatest *vertical distance* (height) you can send the pancake? _____ cm

What angle resulted in the greatest height? _____ °

Sketch this set-up, showing the angle, height reached, and the pancake's flight path.

13. Fold the index card so that it has a base, and stand it about a half meter from the end of the catapult. Find and describe two different ways of hitting the index card target.

_____

_____

_____

_____

14. Place the index card target about half a meter in front of the catapult and practice hitting the target. Try making your target smaller. How accurate can you make your catapult? What could you do to increase the accuracy?

_____

_____

_____

_____

15. Stick two standard pancakes together like the halves of a hamburger bun. What effect does doubling the mass of the projectile have on the distance it travels?

_____

_____

_____

16. Summarize your observations and discoveries by writing some *when-then* statements. For example, *"When* the angle of the catapult is increased, *then. . . ."*

_____

_____

_____

_____

_____

_____

_____

_____

GUIDE TO ACTIVITY 4

Pancake catapults

◆What is happening?

The catapults used in this activity provide a simple and inexpensive way for students to have direct experience with projectile pathways and some factors that influence them.

Even at the modest speeds that a single rubber band produces, some aspects of the pancake's flight are difficult to observe without resorting to observational aids such as strobe photographs or radar guns. For example, it is hard to judge whether or not the pancake continues to accelerate in an upward direction after it leaves the arm of the catapult. It is equally hard to observe the exact shape of the path it follows and the pancake's speed at various points along the path by just watching with the unaided eye.

Because of these observational difficulties, people observing a moving object may believe that the object moved in accordance with their preconceived ideas about motion, when these ideas are in fact incorrect. Their expectations influence what they perceive. Rather than having observations correct misconceptions, misconceptions may only reinforce themselves by contributing to faulty observations. An understanding of the laws of motion is essential in producing logical (and correct) answers to problems like the ones in this activity.

Newton's first law of motion states: An object at rest tends to stay at rest, and an object in motion tends to stay in motion in a straight line and at a constant speed unless acted upon by an unequal force. This means that we don't need causes to *preserve* motion. We need causes to *change* motion.

The pancake acquires its upward motion by interacting with the arm of the catapult. After the pancake leaves the arm, the arm no longer exerts an upward force on the pancake. This force is not required to *maintain* a straight-line, constant-speed motion for the pancake, but this motion *does change* as the pancake flies through the air as a result of two other forces that do act on the pancake: the downward force of gravity and the force of air friction.

◆Time management

Two class periods (40–60 minutes each) may be required to complete the activity and discuss the results. You may wish to do Part III as a separate activity.

◆Preparation

This activity has the potential to create a lot of excitement. Accordingly, careful consideration of safety issues and classroom management is strongly recommended. Do not allow students to modify the catapult design or the pancakes, and make sure that *goggles are worn at all times*.

The clay pancakes have several advantages as projectiles: They are soft and therefore relatively safe; they stick slightly to the popsicle sticks so that the catapults can be raised to different angles; and they are flat and so tend to travel only a short distance after landing.

For younger classes, you may wish to assemble the catapults in advance, go over the safety rules and operating procedures, and allow students to begin with Part II of the procedure.

◆Answers

9. While each individual catapult should perform fairly consistently, the results can vary greatly from catapult to catapult. Generally, distances shot will fall somewhere between 25 cm and 65 cm.

10. Results will vary, but the average value will probably be in the 25–65 cm range.

11. As the angle of the catapult is increased to some middle angle (30–60°), the range of the projectile *increases*. Increasing the angle past this middle angle *decreases* the range. Theoretically, 45° should be that middle angle and should produce the maximum range.

If the strength of the rubber band could be increased, the pancake would fly further and higher since a greater force could be exerted on it. A greater force (F) exerted on the same mass (m—the standard pancake) produces a greater acceleration (a), as follows from the second law of motion, $F = ma$.

12. The greatest height of the pancake should be observed when the stick launches the pancake *straight up*. This *may not* correspond to a measured angle of 90° between the base of the catapult and the book. The critical factor is the angle at which the pancake leaves the arm—not the position of the base.

Getting an accurate determination of the apex is a little tricky. Have observers stand on both sides of the catapult, while a third person holds the meter stick near the expected path of the pancake.

13 & 14. You can hit the target by lobbing it in a high trajectory, or shooting straight at it using a low trajectory. Since the target is placed at about the midpoint of the catapult's range for this challenge, shooting straight is the most accurate approach. (The closer the target is to the end of the catapult's range, the more curved will be the path of the pancake near the target.)

Accuracy depends on using consistent techniques of holding, aiming, and releasing the stick. Up to a certain point, adding power to the catapult by adding extra rubber bands would increase accuracy, too. Extra rubber bands increase the speed of the pancake. This increases its range, thus increasing the distance over which it is possible to shoot straight at the target.

15. When more clay is used in the pancake, the pancake doesn't travel as far or as high.

When more clay is added, the mass (m) of the projectile increases. With more mass and the *same force*—the force applied (F) is determined by the rubber band and does not change—the acceleration (a) decreases (Newton's second law, $F = ma$).

16. Answers will vary for this challenge, but the *when-then* statements should summarize the answers for 11–15. For the example provided, one might write: "*When* the angle of the catapult is increased, *then* the height reached by the pancake and the distance traveled increases."

This *when-then* statement is true for angles of launch up to some middle angle, (theoretically 45°). As the angle is increased beyond 45°, the height increases, but the distance traveled by the pancake decreases. Maximum height (and *minimum* distance) is achieved by firing the catapult straight up—a launch angle of 90°.

MODULE 2

It's All in a Day's Work

◆Introduction

• Who is doing more work: a child picking flowers or a weightlifter straining every muscle but *failing* to raise a barbell off the floor?

• Which requires more work: *rolling* a refrigerator 1 m across a level floor, or *lifting* it 1 m?

• The force of gravity keeps the moon orbiting around the Earth, but Earth's gravity does *no work* to maintain the moon's orbit. How is this possible?

This module examines the physical concept of work. Students will learn the difference between the common and scientific meanings of work; they will also learn to identify the forces involved, why the *direction* of motion is important, and the role gravity plays.

◆Instructional Objectives

After completing the activities and readings for Module 2, students should be able to

• define work in scientific terms [Activity 5]

• explain how friction affects the amount of work being done [Activity 6]

• demonstrate work being done on objects that are moving horizontally, vertically, and at an angle relative to the Earth [Activity 7]

• demonstrate how changing the direction of a force changes the distance that an object moves [Activity 7]

• explain how increasing an object's weight affects the work done raising it [Activity 8]

• calculate work in joules [Activities 5–8]

◆Preparation

Study the following reading for Module 2:

Reading 5: How Hard Are You Working? What Are You Working Against?

◆Activities

This module includes the following activities:

Activity 5: *Eureka!* #8—Work

Activity 6: Work Is a Drag

Activity 7: Working an Angle

Activity 8: Working Against Gravity

ACTIVITY 5: VIDEOTAPE

Eureka! #8—Work

◆Background

Eureka!, produced by TVOntario, is a series of animated videos using examples drawn from everyday experience to demonstrate the behavior of matter in motion. The presentations are an enjoyable way to review the concepts of mechanics. You may wish to use them in the classroom to help students develop an intuitive feel for the principles being studied.

This segment of *Eureka!* defines **work** with the equation

Work = Force x distance.

Examples of a clown and a strongman moving different objects illustrate how work is measured and are used to define the **joule (J)**, the preferred unit in which work is measured.

One joule is equal to a force of 1 newton moving through a distance of 1 meter. This definition can be shown as the equation

1 joule = (1 newton) x (1 meter) = 1 newton meter

or, in metric symbols

$1 J = 1 N \times 1 m = 1 N \cdot m$.

◆Time management

The running time of the videotape is five minutes. At least 15 minutes should be allotted to introduce, run, and discuss the videotape. You may wish to play the videotape at the end of a lesson to reinforce the concepts presented.

◆Comments on the videotape

The videotape makes an important point about the *scientific* definition of work (as opposed to the layman's definition) when it shows the strongman pushing against his car but being unable to move it even one millimeter. People who have not studied physics would *incorrectly* say that the strongman is working very hard while trying to move his car.

A *scientist* watching the strongman's efforts would disagree, saying "Since the car *does not move*, the strongman is doing *no* work. He may be applying a large force to the car, but the distance moved is zero. So, applying the work equation

Work = Force x distance = Force x 0 m = 0 J,

we see that no work is being done."

Concept summary

"Stationary things don't want to move—it takes a force to move them."*

"Force equals mass times acceleration and is measured in newtons."*

"When a force moves something through a distance, work is done."*

"Work is measured in joules."*

"One joule is the amount of work done when 1 newton of force is applied through a distance of 1 meter."*

Eureka! Produced by TVOntario ©1981.

ACTIVITY 6 WORKSHEET

Work Is a Drag

Each group will need
• 50 cm of string
• a common brick (preferably one with holes through it)
• duct tape or masking tape
• a smooth surface to drag the brick across—a lab table or tile or vinyl floor
• a meter stick
• a spring scale calibrated in newtons or grams
• 3 ice cubes

Vocabulary

• **Friction:** Resistance to relative motion between objects in contact. The force due to friction acts on an object in the direction opposite to that of its motion.
• **Joule:** The preferred unit of measurement for work in the metric system. One joule (1 J) is the amount of work done when a force of 1 N is applied through a distance of 1 m.
• **Newton:** The standard metric unit of force. One newton (1 N) is the amount of force required to accelerate a 1000-g mass at a rate of approximately 1 m/s^2.
• **Work:** The product of an applied force and the distance through which it acts.

Note: Check the brick for irregularities that protrude from its surface. If the brick is smooth, it will not harm a formica table top or a tile or vinyl floor covering.

◆Background

Work may be a drag, but probably not in the way you think. When scientists talk about work, they mean something very different from how most of us use the word. **Work**, as a physical concept, is the product of an applied force and the distance through which the force acts. You may think that you are working hard to solve a homework problem, but if you are not applying a force to your papers and making them move, then as far as scientists are concerned you are not doing any work. The equation for work is

　　　Work = Force x distance

or,

　　　W = F x d.

　If we calculate the amount of work done when a force of 1 **newton (N)** is applied through a distance of 1 m, we find

　　　W = F x d = 1 N x 1 m = 1 N • m.

In the metric system, 1 N m is called a **joule (J)**. So, we can write

　　　W = F x d = 1 N x 1 m = 1 N • m = 1 J.

Joules are the preferred units of measurement for work in the metric system.

　Suppose we want to calculate how much work is required to drag a brick a certain distance. Since W = F x d, we would want to know what force is applied *throughout* the move (F), and how far the brick moved (d). Then we could multiply the two together to find the work. The **friction** between the brick and the surface beneath it would produce a force acting opposite to the direction of motion. How would this affect our calculations? In this activity you will determine how much work is required to drag a brick 1 m and explore how friction is involved.

◆Objective

To learn how friction affects the amount of work required to move an object

◆Procedure

1. Tape the string to one end of the brick to form a U-shaped harness.

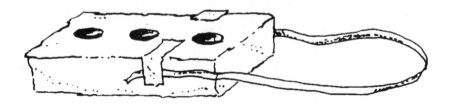

2. Find a smooth surface that will not be damaged when you drag the brick across it.

Stick two small pieces of tape 1 m apart on this smooth surface. These pieces mark the Start and Finish lines.

3. Place the brick 50 cm behind the Start line. Attach the hook of the spring scale to the loop of string forming the harness.

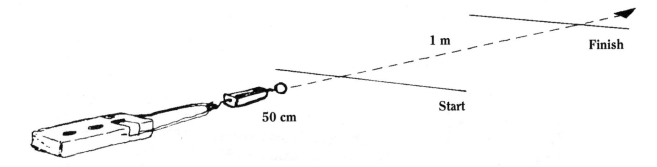

4. Hold the spring scale so that you can read it while pulling the brick down the 1-m course.

Use a smooth, steady pull to set the brick in motion. Once the brick is moving, try to *keep its speed constant*.

5. Read the spring scale when the brick is even with the Start line. This reading indicates how much force (in newtons) is being exerted by the person pulling the brick. Try to keep the brick moving at a constant speed. If its speed is constant, then its acceleration (a) in that direction is zero and Newton's second law, $F = ma$, shows us that the net force (F) acting on the brick in the plane of motion must also be zero. The net force is zero only when the two forces operating in the plane of motion—the pull and the force of friction—are equal and opposite. So keeping the speed constant means that the strength of the pull force, as read on the scale, is equal to the strength of the force due to friction.

6. Read the spring scale again when the brick reaches the Finish line. *Do not stop pulling the brick until you have passed the Finish line*; keep the same constant speed until the brick is beyond the Finish line and you have read the scale.

If you pulled the brick at a constant speed, both force readings should be about the same. If the two force readings differ greatly, repeat steps 4, 5, and 6 until you get similar readings at the Start and Finish lines.

7. Write your force readings in the data table.

Note: If the spring scale is calibrated in grams, divide the reading by 100 to find the force in newtons.

Note: As the ice cubes melt, the course may get slicker, further reducing the force required to pull the brick.

Data table

Force required to move a brick

	Start	Finish	Average force
Brick placed directly on the surface	_____ N	_____ N	_____ N
Brick supported by ice cubes	_____ N	_____ N	_____ N

8. Remember that the force in the definition of work is applied *throughout the entire distance* that an object is moved. In most situations, this force will vary in strength and even direction while being applied. We therefore need to find the *average force* applied in order to calculate work. Calculate the average force required to move the brick by adding together the force readings at the Start and Finish lines and then dividing by two. Enter your result in the data table.

9. Now let's try reducing the force of friction acting on the brick.

Place three ice cubes underneath the brick to support it above the surface. The ice cubes should be placed so that the brick will not tip over when the string harness is pulled.

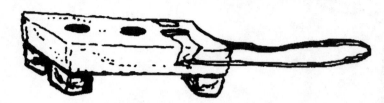

10. Place the brick (supported by the ice cubes) 50 cm behind the Start line. Attach the scale to the harness, and tow the brick down the 1-m course at as close to a constant speed as possible.

Read the force being exerted on the brick supported by ice cubes at the Start and Finish lines in the same way that you did for steps 5 and 6.

The Start and Finish force readings for the brick supported by ice cubes should be very similar. If they differ greatly from one another, pull the brick down the course again, being careful to maintain a constant speed.

11. Enter these force measurements in the data table.

Calculate the average of the two force readings obtained for the brick supported on ice and enter this result in the data table.

12. Use the average force for the brick placed directly on the surface to calculate the work performed while moving the brick 1 m:

W = F x d = _____ N x 1 m = _____ N • m = _____ J.

13. Use the average force for the brick *supported by ice cubes* to calculate the work performed while moving the brick 1 m:

W = F x d = _____ N x 1 m = _____ N • m = _____ J.

14. Now answer the question posed at the beginning of this activity: How does friction affect the amount of work required to move an object?

_____

_____

_____

_____

GUIDE TO ACTIVITY 6

Work Is a Drag

◆What is happening?

This activity demonstrates how friction affects the amount of force required to slide a brick along a smooth horizontal plane. When the force of friction between the brick and the surface is reduced, moving the brick 1 m requires less work because it requires a smaller force.

The friction between the brick and the surface determines the magnitude of the pulling force necessary to keep the brick moving at a constant speed. The pull has to be equal and opposite to the force due to friction to achieve a net force of zero acting on the brick in the plane of motion. Newton's second law, $F = ma$, shows that when the net force (F) is zero, then the acceleration (a) on the mass of the brick is also zero—we therefore have constant speed. A pulling force greater than the force due to friction would result in the brick accelerating (which does happen in the first moments of pulling since the brick increases in speed from zero). Placing ice cubes under the brick reduces the friction between the brick and the surface. (In other words, the melting ice serves as a lubricant.) Decreasing the friction means a smaller pulling force is required to overcome and then balance the force due to friction. A smaller force means less work.

How much work would be required to keep the brick moving at a constant speed along the surface if there were no frictional forces acting on the brick? The answer to this question surprises many people: *No work would be required!* To explain this answer, think back to the first law of motion: An object in motion tends to stay in motion in a straight line and at a constant speed unless acted upon by an unequal force.

Friction is an unequal force that acts on the brick. In the absence of all friction, the brick's inertia would keep it moving along the surface at a constant speed without the need for any friction-countering force being applied to it and therefore without any work being done on the brick. *Accelerating* the brick up to that constant speed would require work, however. The force required to accelerate a given mass is stated in the second law of motion: $F = ma$.

◆Time management

One class period (40–60 minutes) should be enough time to complete the activity and discuss the results.

◆Preparation

Remind students to take force readings *while the brick is still moving at a constant speed*. When the brick is moving at a constant speed, the strength of the force of friction between the brick and the surface is equal to the strength of the pulling force being registered on the scale. This will simplify comparing the effects of friction on work in the two set-ups. Make sure that students keep pulling the brick *through* the Finish line; they should not begin to slow their pull until after they have passed the Finish line.

"Bought" ice cubes are more uniform in shape and size and have more parallel surfaces than "homemade" ice cubes, and therefore make a better support for the brick. As the ice melts, the friction between the ice and the surface will decrease. You may wish to minimize friction further by wetting the surface before placing the ice on it.

Students can use a tile or vinyl floor, lab tables with a hard surface finish, or a piece of unfinished shelving board to drag the bricks across. If you are concerned about damaging the surface, you can tape a piece of cardboard to the brick to cushion it. This will affect the friction between the brick and the surface, but the general relationship between friction and work can still be demonstrated.

If getting ice is a problem, place dowels or round pencils along the 1-m course. These rollers will also reduce the friction between the brick and the surface.

◆Suggestions for further study

Several factors affect the friction present between the brick and the surface, and thus the amount of force required to slide the brick: the weight of the brick, the composition of the surface on which it is sliding, and the presence (or absence) of a lubricant (such as ice and water) between the brick and the surface. Theoretically, the amount of surface area of the brick that is in contact with the sliding surface does not affect the friction between the brick and the surface. Have your students design experiments that test the effects (or absence of effects) of these variables on the amount of work required to move the brick 1 m.

◆Answers

The exact results will vary depending on the size and roughness of the brick and the surface being used. The following sample data are included *only as an example* to suggest the range of values that you may find. *These data are not the only correct answers.*

Average force for brick placed directly on a pine board: 7 N

Average force for brick supported on ice on a pine board: 2 N

Average force for brick placed directly on a lab table: 3 N

Average force for brick supported on ice on a wet lab table: 0.5 N

12. *Answers will vary.* The following is one possible result:
W = 7 N x 1 m = 7 N • m = 7 J.

13. *Answers will vary.* The following is one possible result:
W = 2 N x 1 m = 2 N • m = 2 J.

14. When an object is moving horizontally along a level surface, the friction between the object and the surface determines how much work is required to move the object at a constant speed. The more friction, the more work is required; the less friction, the less work is required.

ACTIVITY 7 WORKSHEET

Working an Angle

◆Background

Why can you coast farther on your bike on a level stretch than you can up a hill? You may pedal just as hard before starting to coast, and gravity pulls you and your bike down just as hard, but you don't travel as far up the hill. Why? When asking questions like these about force and distance, we often forget that there are *two* properties to a force: strength *and* direction. How does the direction of a force affect situations like this, and how does it affect the concept of work, which is the product of force and distance?

◆Objective

To explore the importance of a force's direction when calculating work

◆Procedure

1. Place the section of track on a meter stick. Straighten any curves in the track, then tape both ends of the track to the meter stick as shown in the diagram. Do not place any tape on the middle of the track.

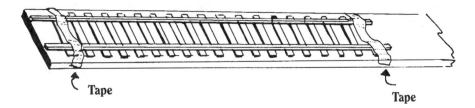

Tape   Tape

2. Tape the *large tube* of the force producer at one end of the track. The tape should be wrapped firmly enough to hold the large tube in place, but *not so tightly* that the small tube cannot slide freely along the track. If the small tube binds against the track, loosen the tape.

Small tube must slide freely

Direction of force

Tape force producer to track

3. Test the assembled apparatus as follows:
• *Put on your safety glasses.*
• Place the meter stick with the track on a flat surface.
• Pull the small tube of the force producer to the medium force mark.
• Place the marble so that it is just touching the small tube.
• Release the small tube.
The marble should roll along the entire length of track, then drop off the end.

Materials

Each group will need
• safety glasses or goggles (for each member of the group)
• a force producer (to be assembled before class by the teacher—see the Preparation section of the Guide to Activity 7)
• a large marble (a "shooter" approximately 2.5 cm in diameter)
• 2 meter sticks
• a section of N-gauge model railroad track approximately 75 cm long
• masking tape
• several books (or other objects to hold the track up at an angle)
• a protractor

 Everyone should wear safety glasses or goggles while setting up and performing this activity!

 Do not point the force producer at anyone! The small tube is not attached to the large tube and could therefore fly off.

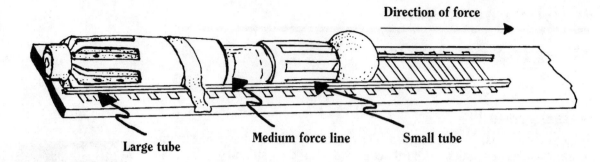

Direction of force

Large tube   **Medium force line**   **Small tube**

If the small tube does not push the marble smoothly, loosen the tape holding down the large tube.

If the medium force position of the tube is not strong enough to push the marble off the end of the track, pull the small tube farther back toward the large tube and mark a new medium force position.

If the medium force position of the tube shoots the marble too far, mark a lower force position on the tube.

4. Support the meter stick and track on a stack of books so that the track is at a 45° angle to the table. Use a protractor to check the angle between the board and the table.

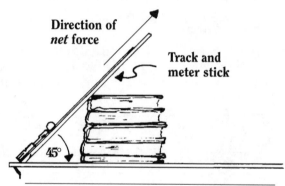

Direction of net force

Track and meter stick

45°

5. Let the force producer come to rest at the zero force position.

Place a small piece of masking tape beside the track to mark the location of the tip of the small tube relative to the track when set at the zero force position.

6. Pull the small tube of the force producer to the medium force mark, and place the marble against the tip of the small tube. Release the small tube to launch the marble.

Measure the distance between the zero force tape marker and the highest point the marble reaches on the track. Record this distance in the data table that follows as Trial 1 for the 45° angle.

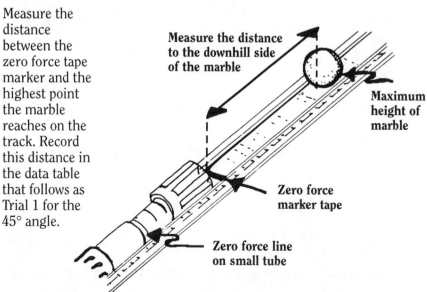

Measure the distance to the downhill side of the marble

Maximum height of marble

Zero force marker tape

Zero force line on small tube

Distance the marble travels along the inclined track

	Trial 1	Trial 2	Trial 3	Trial 4	Average distance
45° angle of launch	_____ cm	_____ cm	_____ cm	_____ cm	_____ cm
15° angle of launch	_____ cm	_____ cm	_____ cm	_____ cm	_____ cm
5° angle of launch	_____ cm	_____ cm	_____ cm	_____ cm	_____ cm

7. With the track at a 45° angle, launch the marble three more times and record your measurements for each of these trials in the data table. Average the 4 trials, and enter the result in the data table.

8. When you have completed four trials at a 45° angle, lower the track to a 15° angle. Using the procedure described in step 6, launch the marble four times at a 15° angle and measure the distance that the marble travels.

Record your observations in the data table, then compute the average of the four trials.

9. When you have completed 4 trials at a 15° angle, lower the track to a 5° angle, repeat step 6, and enter the measurements for the 4 trials in the data table. Compute the average of the 4 trials.

10. Which angle of the track (45°, 15°, or 5°) allowed the marble to travel the *longest* distance? _____ °
Which angle of the track (45°, 15°, or 5°) allowed the marble to travel the *shortest* distance? _____ °

11. Which angle of the track (45°, 15°, or 5°) allows the force of gravity to have the *greatest* effect on the distance that the marble moves? _____ °

12. Why is it important to find the *direction* of the force *before* using the formula *Work = Force x distance* to compare the work done on an object?

_____

_____

_____

_____

GUIDE TO ACTIVITY 7

Working an Angle

◆What is happening?

Uncritically applying the formula W = F x d can lead to erroneous conclusions about how much work is being done on an object.

The force producer used in this activity applies the same amount of force to the marble for each trial. The force applied to the marble gives it about the same initial speed for each trial. However, the total *distance* the marble moves changes when the *direction* of the net force acting on the marble is changed.

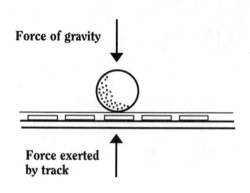

Force of gravity

Force exerted by track

When the marble is sitting still on a level track, the force of gravity exerts a *downward* force on the marble, and the track exerts an *upward* force on the marble. Because these forces are *equal* in magnitude and *opposite* in direction on a level track, this pair of forces cancels out. The marble remains at rest because there is no net force acting on it.

Two additional forces act on the marble when the force producer is used to set it into motion: the forward push of the force producer and the backward push of friction. The forward push is much larger than the force of friction, so the horizontal push overcomes the ball's inertia and the marble moves along the track.

The force producer cannot affect the motion of the marble after the marble loses contact with the force producer's tube, because the force producer no longer exerts a force on the marble. Once it is set in motion, the marble continues

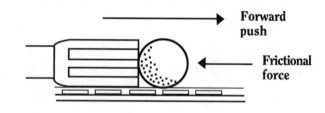

Forward push

Frictional force

in motion in a straight line because of its inertia (force is necessary to *change* motion, but not to *maintain* motion).

Newton's first law predicts that a marble moving in a complete vacuum along a perfectly straight, zero-friction track would continue moving at a constant speed until an unequal force acted on it. Since it is impossible to eliminate completely the unequal force of friction between the marble and the track, marbles in the real world always stop moving.

Friction is a hidden force (hidden in the sense that we sometimes overlook it) that acts to bring the marble to a halt. The downward force of gravity acting on the marble *does not slow the marble*, because when the track is horizontal to the Earth, the force of gravity is canceled out by the equal and opposite (upward) force exerted on the marble by the track.

When we tilt the track, the force that gravity exerts on the marble and the force that the track exerts

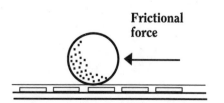

Frictional force

on the marble no longer line up. The force of gravity on the marble still acts straight down, but the force of the track on the marble, since it is always perpendicular to the surface of the track, is no longer acting straight up. We can see in the following diagram, however, that the force of the track acting on the marble (F_T) still counters part of the force of gravity acting on the marble (F_G). We can consider F_G as the sum of two forces: one that acts parallel to the track (F_{GP}) and one that acts perpendicular (normal) to the track (F_{GN}). Experience shows you that while the marble rolls down the track, it neither lifts off or falls through the track. This is because the force of the track (F_T) is equal and opposite to the part of the gravity force that is trying to pull the marble straight through the track (F_{GN}): F_T cancels out F_{GN}. But the force of the track does not affect the part of the force of gravity that acts downward along the track (F_{GP}). So, tilting the track causes gravity to exert an unbalanced force downward along the track.

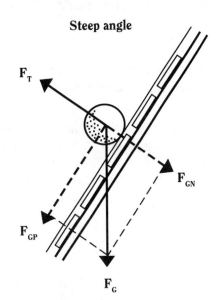

Steep angle

The force producer creates a force that acts parallel to the track, no matter the position of the track. When the track is level, only the force due to friction opposes the force producer's force. But tilting the track causes gravity to exert a force that acts opposite to the force producer's force. In fact, the steeper the track, the stronger the force gravity applies along the track. So, as the angle increases, the *net* force pushing the marble forward along the track decreases, and as a result, the distance the ball travels decreases (by Newton's second law, a smaller force acting on the same mass produces a smaller acceleration). If we did not consider the angles at which all these forces are interacting, we would not understand why the ball does not travel as far at a steeper angle. We might think that of the two quantities we use to calculate work only one, distance, is changing—but now we see that the other, the net force acting on the marble, is changing too. Gravity pulls just as hard, the

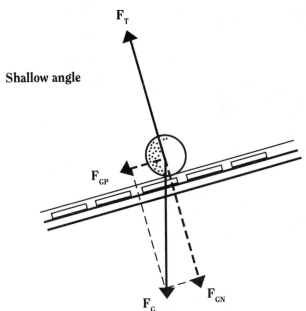

Shallow angle

force producer pushes just as hard, but the net force *changes*—all because the *angle* changes. We could never calculate work without being aware of the angles at which the various forces operate.

◆Time management

One class period (40–60 minutes) should be enough time to complete the activity and discuss the results.

◆Preparation

Both this activity and Activity 8 require a device that can be relied upon to produce a constant force. Following are assembly instructions for an inexpensive, convenient device for generating a reproducible amount of force. This device is just one example of how everyday items can be used or modified for use in laboratory activities.

Assembly instructions for a device that produces a constant force

1. For each device you will need

• a Handi-Mates™ replacement tissue roller #02511 by Mirra-Cote™ (available at K-Mart™ stores) or a similar plastic roller from another manufacturer

• a fine-toothed handsaw (a coping saw or hacksaw will work well)

• a small piece of medium-grit sandpaper

• a drill with 1/8" bit

• a fine-point permanent black marking pen

• safety glasses

• **optional:** a paper clip and candle (instead of the drill)

2. Put on your safety glasses. Wear them at all times while assembling or testing the force producer.

3. The roller used for this activity consists of three parts: a large-diameter outer tube, a smaller tube that fits inside the outer tube, and a spring that fits between the two tubes.

Separate the small and large tubes. If the spring comes out, press it back into the large tube.

Use the saw to cut off the tip of the small tube.

Hold the tube so that your fingers will not be injured if the saw slips.

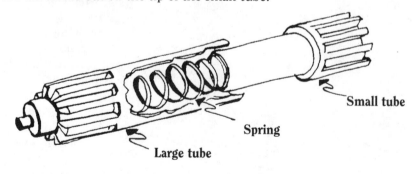

Small tube

Spring

Large tube

4. Sand any rough edges produced by the cut. (Just level the roughest edges—the cut does not need to be perfectly smooth.)

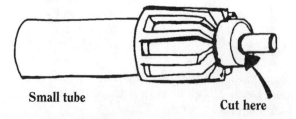

Small tube

Cut here

5. Drill two holes in each of the grooves at the end of the large tube. Do not drill any holes in the smooth barrel of the large tube, or

the small tube will not slide through the large tube smoothly. If you do not wish to use a drill, an alternative procedure is to heat the end of a paper clip over a candle and push the hot end through the roller. The plastic is

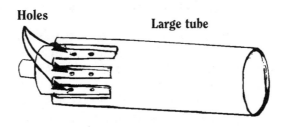

Holes

Large tube

very soft, allowing this method to work quite easily. You may wish to hold the paper clip with a pair of pliers to avoid burns.

6. Replace the small tube inside the larger tube. Press the small tube down on the spring to be sure that the small tube slides back and forth smoothly.

If the spring is loose or is not seated at the end of the large tube, turn the spring like a corkscrew inside the large tube until it touches the bottom of the tube.

7. Calibrating the assembled force producer

• Finding the *zero force* position:

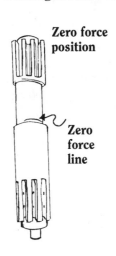

Zero force position

Zero force line

Place the end of the large tube on the table. Hold the force producer vertically and allow the small tube to rest on the spring. Use a sharp-tipped marker to draw a line on the small tube where it meets the large tube. Write *0 F* beside this line.

• Finding the *maximum force* position:

Place the end of the large tube on the table. Hold the force producer vertically as before. Press the small tube into the large tube until the spring cannot be compressed any further. Use a sharp-tipped marker to draw a line on the small tube where it meets the large tube. Write *Max F* beside this line.

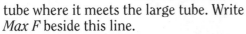

Maximum force position

Maximum force line

• Estimating the *medium force* position:

Remove the small tube from the large tube. Use a sharp-tipped marker to draw a line midway between the zero force and maximum force lines. Write *Med F* beside this line.

The force producer is now complete.

If you do not know how to operate a drill safely, please have an experienced woodworker demonstrate the proper way to use one. If you purchase a drill for this activity, study the safety procedures in the owner's manual.

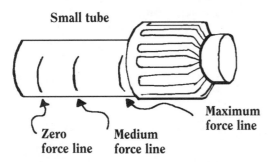

Small tube

Zero force line   **Medium force line**   **Maximum force line**

The modifications to the roller improve its usefulness as a force producer. Cutting off the tip of the tube produces a flat surface that allows the force of the spring to be transferred to other objects more efficiently.

A partial vacuum forms when the spring pushes the small tube out of the large tube. The holes allow air to rush into the large tube and equalize the pressure. Without the holes, the small tube would be held back by the unequal air pressure.

The following suggestions may improve the performance of the force producer/track apparatus, and make data collection easier:

• Place the track on the meter stick in such a way that the distance markings on the stick can still be read. This will eliminate the need for two meter sticks per lab group.

• If you tape the track to the meter stick in this way, you should also tape the force producer on the track so that when it is in the zero force position, the *tip* of the small tube is *even with the 10-cm or 20-cm mark* on the meter stick. Using this arrangement, students can easily determine the distance that the marble moves by subtracting 10 cm or 20 cm from the maximum height that they read directly from the meter stick supporting the track.

◆Suggestions for further study

Look at the data you have recorded and predict the average distance the marble should move up the track when the angle is 10°, 30°, and 60°. Test your predictions.

Assume you had the job of rolling a snowball to the top of a hill. You could take a route that is long but not steep (small angle), or a route that is short but steep (large angle). Which route would require that you exert a greater force on the snowball? Which route would require that you exert that force over a shorter distance? Recall that W = F x d. Compare the work done on the short and long routes.

◆Answers

Data: The actual data that you obtain will vary depending on the strength of the spring, the position of the medium force mark, the release technique, and the mass of the marble. The following observations are included for *reference only. These answers are not the only correct and acceptable answers for this activity.*

Average distance for 45° angle of launch: 9 cm

Average distance for 15° angle of launch: 25 cm

Average distance for 5° angle of launch: 35 cm

10. The marble travels the longest distance at a 5° angle of launch.

The marble travels the shortest distance at a 45° angle of launch.

11. Gravity has the greatest effect on the marble launched at a 45° angle.

12. The force producer applies the same amount of force to the marble for each trial, no matter what direction the track is pointing. However, when the track is placed at an angle, gravity exerts a force on the marble that opposes the force that the force producer exerts on the marble. Therefore, the *net* force forward along the track is less. So, even though the force of the force producer never changes, the change in *direction* makes the net force on the marble (*F* in W = F x d) less than it is when the track is level.

ACTIVITY 8 WORKSHEET

Working Against Gravity

◆Background

Suppose a hiker in Yellowstone Park inadvertently disturbs a grizzly bear. The bear charges, and the hiker tries to reach safety by leaping into a tree. Could the hiker leap higher while wearing a heavy backpack, or without the pack? The hiker exerts a force when he or she leaps. The hiker's weight and the weight of the backpack are also forces. We are curious about how the height of the leap—a distance—is affected by these forces. Work is the product of force and distance. Can we apply the concept of work to this situation? What can we learn about work?

◆Objective

To explore the relationship between force, distance, and weight with the concept of work

◆Procedure

1. Determine the mass of the small (inside) tube of the force producer with a laboratory balance.

Mass of tube = _____ g

2. Place one of the adhesive balls on the balance. Add or pull off bits of adhesive until the ball weighs *about* the same as the small tube. (If the tube has a mass of 10 g, the mass of the ball should be adjusted to between 9.5 g and 10.5 g.) Use the same procedure to adjust the second ball so that its mass also equals the mass of the small tube.

3. **Put on your safety glasses before proceeding.** Do not point the force producer at yourself or anyone else while performing the procedure.

4. Have one member of the group hold the meter stick vertically in the middle of the table. Turn the stick so that the 1-cm mark is near the table top.

5. Place the large tube of the force producer on the table about 5 cm in front of the meter stick. Point the small tube *straight up*.

The person holding the force producer should press the small tube down to the maximum force mark. Then quickly and smoothly release the small tube so that it flies straight up, parallel to the meter stick.

Materials

Each group will need

• safety glasses for each person

• the force producer from Activity 7 (assembly instructions can be found in the Guide to Activity 7)

• a meter stick

• Eberhard Faber Holdit™ plastic adhesive—two grape-size balls

• a laboratory balance calibrated in grams

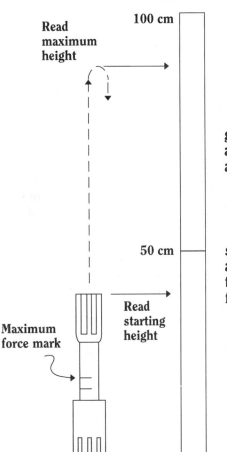

Meter stick

100 cm

Read maximum height

50 cm

Read starting height

Maximum force mark

1 cm

Everyone should wear safety glasses or goggles while setting up and performing this activity!

Do not point the force producer at anyone! The small tube is not attached to the large tube and could therefore fly off.

6. Practice launching the tube until each flight is almost vertical and the tube reaches about the same height each time. Use the meter stick to judge the height reached by the top of the tube.

Be careful not to block the holes in the large tube. If the spring pops loose, turn the spring like a corkscrew until it stops at the bottom of the tube. When you feel confident that you can release the tube so as to produce consistent flights, go on to Step 7.

7. Place the force producer next to the meter stick and push the small tube down against the spring until it stops.

Determine the height of the top of the small tube, and record it below.

Starting height = _____ cm

8. Launch the tube five times. For each launch, record the *maximum height* that the top of the tube reaches in the data table.

Data table

Maximum height

	Small tube	Tube + 1 ball	Tube + 2 balls
	_____ cm	_____ cm	_____ cm
	_____ cm	_____ cm	_____ cm
	_____ cm	_____ cm	_____ cm
	_____ cm	_____ cm	_____ cm
	_____ cm	_____ cm	_____ cm
Average maximum height	_____ cm	_____ cm	_____ cm

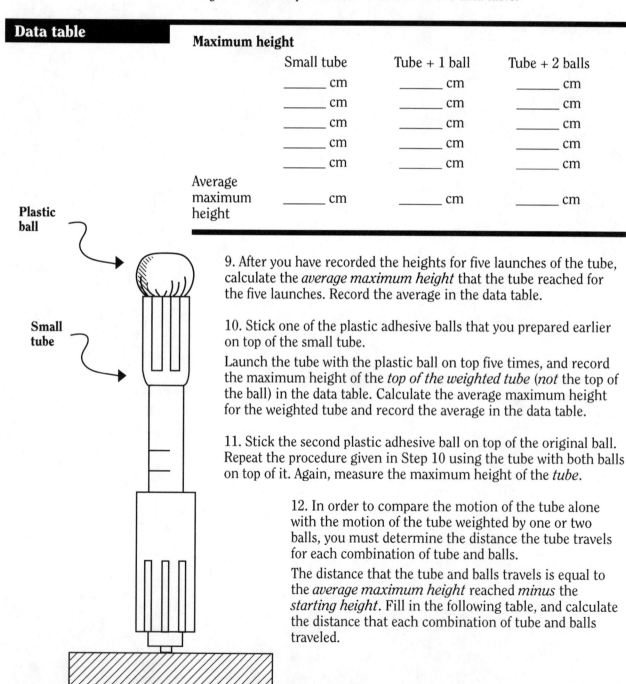

Plastic ball

Small tube

9. After you have recorded the heights for five launches of the tube, calculate the *average maximum height* that the tube reached for the five launches. Record the average in the data table.

10. Stick one of the plastic adhesive balls that you prepared earlier on top of the small tube.

Launch the tube with the plastic ball on top five times, and record the maximum height of the *top of the weighted tube* (*not* the top of the ball) in the data table. Calculate the average maximum height for the weighted tube and record the average in the data table.

11. Stick the second plastic adhesive ball on top of the original ball. Repeat the procedure given in Step 10 using the tube with both balls on top of it. Again, measure the maximum height of the *tube*.

12. In order to compare the motion of the tube alone with the motion of the tube weighted by one or two balls, you must determine the distance the tube travels for each combination of tube and balls.

The distance that the tube and balls travels is equal to the *average maximum height* reached *minus* the *starting height*. Fill in the following table, and calculate the distance that each combination of tube and balls traveled.

Calculating the distance traveled by the tube and balls

Object being tested	Average maximum height	–	Starting height	=	Distance traveled (cm)
Tube alone	_____ cm	–	_____ cm	=	_____ cm
Tube + 1 ball	_____ cm	–	_____ cm	=	_____ cm
Tube + 2 balls	_____ cm	–	_____ cm	=	_____ cm

13. Which of the three objects traveled the greatest distance?

_____

14. Which of the three objects traveled the shortest distance?

_____

15. Which of the three objects is heaviest?

_____

16. Which of the three objects is lightest?

_____

17. Write a statement that summarizes the general relationship between the *weight* of an object and the *distance* it travels upward when it is acted upon by a constant amount of force.

_____

_____

_____

_____

18. Directly calculating the work done on the three objects by the spring would be very difficult. For instance, the distance term in the work equation (Work = Force x distance) is equal to only the very small distance the objects travel *while still in contact with the force producer.* (Remember: Work is only being done while the force is being applied; once the objects are no longer in contact with the force producer, the spring is no longer exerting a force on them.) It would be very hard to measure this small distance accurately. Also, the force produced by the spring *changes* as the spring lengthens during the short time the force producer is in contact with the objects. So how can we calculate the work done on the objects?

Let's consider the work that *gravity* does on the objects. The force in the work equation would then be the force that gravity exerts on the object—its weight. The distance would then be the distance the object moves in the gravity field—the distance you measured. So, we can easily calculate work this way. But how does this compare to the work done on the object by the spring? It turns out to be the *same*. Once you and your teacher have discussed energy and energy conversions you will see why this is so.

Calculate the work done on each of the three objects with the method just described. (Remember that the weight of the objects has to be expressed in newtons. Take the mass in grams and divide by 100 to convert to newtons). The distance traveled should be expressed in meters, so you will also have to divide your measurements in centimeters by 100.

Comparison of work done

Object	Force (weight)	x	distance	=	Work
Tube alone	_____ N	x	_____ m	=	_____ J
Tube + 1 ball	_____ N	x	_____ m	=	_____ J
Tube + 2 balls	_____ N	x	_____ m	=	_____ J

As the weight of the object being launched increases, what happens to the work being done?

_____

_____

_____

_____

19. What are some of the possible sources of experimental error in measurement for this activity?

_____

_____

_____

_____

_____

_____

GUIDE TO ACTIVITY 8

Working Against Gravity

◆What is happening?

Most experimenters will find that as the *weight* placed on the spring *increases*, the *distance* that the spring moves the load *decreases*. We can use these observations to make some inferences about the force exerted by the spring, and about how much work the spring is doing.

At the maximum force position, the spring cannot be further compressed. Therefore, the total amount of force that the spring exerts while in the maximum force position remains constant, no matter how much weight is placed upon it.

You observed that the heavier the tube, the smaller the distance that the spring can lift it. *Doubling* the weight of the tube (by placing a ball on it) approximately *halves* the height that the tube reaches. When the weight of the tube is tripled (by sticking two balls on it), the tube will rise to only about one-third the height reached by the tube alone. These data suggest that the height to which the spring can lift the tube is *inversely proportional* to the weight of the tube.

Everybody knows that it requires more work to lift a heavy object than to lift a light object. In everyday life, we rarely encounter situations in which the force being exerted on objects of different weights remains constant.

Consider the following example. When lifting a heavy bag of groceries (weighing about 100 N, or a little over 20 lbs) from the floor to a countertop about 1 m high, the muscles of the person lifting the bag produce an upward force of 100 N over a distance of 1 m. What might happen if the next bag of groceries to be lifted weighed only 30 N? Would the person still exert 100 N of force on the light bag? Probably not; if the 100 N of force were applied to the light bag over a distance of 1 m, when released the bag would fly up and hit the ceiling.

When our muscles supply the force to move an object, we unconsciously increase or decrease the amount of force being exerted in order to accomplish the desired task. Variable force is not an issue when a spring is compressed the same amount each time before being released; the force that the spring exerts remains constant. This allows us to observe how distance varies as the weight of the object being moved varies.

Directly determining how much work the spring is doing in moving the tube or the tube plus the balls is a tricky task. As the spring pushes the tube upward, it changes length. As it changes length, the amount of force it exerts at any given moment also changes. When it is fully compressed (at the maximum force line), the spring exerts its strongest force. When it lengthens almost to the zero force line, the spring exerts a very small amount of force. The spring pushes against the tube in the same way for each trial, but determining a single numerical value to use as the spring's force in order to calculate work is difficult.

The last term in the equation for work is *distance*. The distance that must be used to calculate the work done by the spring is the distance that the tube moves *while it is touching the spring—not* the total distance that the tube moves. Why is this true?

The spring of the force producer is doing work *only* while it is touching the tube. As soon as the tube loses contact with the spring, the spring exerts no force on the tube; therefore, it can do *no more work* on the tube. The tube continues moving upward against the force of gravity because of its inertia. (Recall the first law of motion: An object at rest tends to stay at rest, and *an object in motion tends to stay in motion* in a straight line and at a constant speed unless acted upon by an unequal force.)

As you see, directly calculating the work done by this apparatus is somewhat complicated. However, it is possible to use the data collected to get a rough *estimate* of the amount of work done by the force producer. The following is a discussion of the assumptions and simplifications involved in making estimates of the work done.

Think of the force producer as a device that transfers "the ability to do work" to the tube and balls. This is another way of saying that the force producer is an *energy source*. Energy is defined as the ability to do work. (Module 3 introduces and explores the concepts of energy and energy conversions.)

If a "perfect" force producer could be built, it would not be affected by friction. It would launch a tube and ball straight up in such a way that the tube would rise to a maximum height, then fall back, landing on the spring. The weight (downward force) of the tube would compress the spring, and the tube would be relaunched, reaching the same maximum height again. This perfect force producer would be a perpetual motion machine that could:

• transfer 100 percent of its ability to do work (energy) to the tube and ball, and

• recapture 100 percent of the energy that it stored in the tube and ball each time the tube falls back onto its spring.

The cycle of rising and falling back on the spring would continue forever in a perfect force producer. In each cycle, the tube of this perfect force producer would stop rising at its maximum height because of the force of gravity doing work on the tube. Gravity would cause the tube to decelerate, stop for an instant, then accelerate back toward the spring of the force producer.

Since this *imaginary* force producer is frictionless and perfect, making the tube rise takes *exactly the same amount of work* as *stopping* the tube from rising. The same quantity of energy is recovered by the perfect force producer each time the tube falls on its spring, compressing it.

The spring of the perfect force producer does the same amount of work in each cycle. The amount of work done by the spring is equivalent to the *gravitational potential energy* (see Module 3) of the tube at the top of its rise—which is equal to the tube's weight times the distance that the tube rises.

Building a perfect machine is impossible, but the real force producer that you built behaves in much the same way, *except for the effects of friction*. Because of friction, the work done by the spring of your force producer is *not exactly equal* to the work done by the force of gravity. However, multiplying the *weight* of the tube (or the tube plus balls) by the *distance* that the tube rises (its maximum height) gives a numerical value that is *proportional to the work done by the spring*. Although it is not perfectly accurate, this value for work is a useful estimate; it avoids the problem of deciding how much force the spring is really applying to the tube.

Making consistent, precise measurements using the simple force producer is difficult. A certain amount of error in measurement is unavoidable, and this experimental error may make interpreting your results difficult. However, if you use good experimental techniques, your data should support the following conclusion: *The amount of work performed remains constant as the weight on the tube is increased.*

◆Time management

One class period (40–60 minutes) should be enough time to complete the activity and discuss the results.

◆Preparation

Make sure that all students wear safety glasses while performing this activity. The force producer should never be aimed at another person.

Encourage students to practice using the force producers before collecting data for this activity. The data will be more consistent if the force producer is released uniformly each time.

Plastic adhesives other than Holdit™, such as Handi-Tak™, may be used to make the balls. Plastic modeling clay may also be used, but clay balls are more likely to come off the tube during the launch.

◆Suggestions for further study

Perform a similar experiment using the force producer, but investigate the maximum height achieved when the *force varies* and the *weight* of the tube remains constant.

◆Answers

Note: These data represent only *one possible set of observations*. The data and answers that follow are intended to serve as a guideline. They are *not* the only acceptable answers for this activity.

1. The mass of the tube will be *approximately* 12 g.

7. The starting height will be *approximately* 11 cm.

Sample data

Maximum height

Tube alone: Average maximum height = 76 cm
(*individual data points range from 73 cm to 80 cm*)

Tube + 1 ball: Average maximum height = 43 cm
(*individual data points range from 42 cm to 45 cm*)

Tube + 2 balls: Average maximum height = 31 cm
(*individual data points range from 30 cm to 32 cm*)

Distance traveled

Tube alone: 76 cm - 11 cm = 65 cm

Tube + 1 ball: 43 cm - 11 cm = 32 cm

Tube + 2 balls: 31 cm - 11 cm = 20 cm

---

13. The tube alone travels the greatest distance.

14. The tube plus two balls travels the least distance.

15. The tube plus two balls is the heaviest.

16. The tube alone is the lightest.

17. When acted upon by an upward force, the lighter the object, the greater the distance it will travel.

18. Comparison of work done:

Tube alone: 0.12 N x 0.65 m = 0.1 J

Tube + 1 ball: 0.24 N x 0.32 m = 0.1 J

Tube + 2 balls: 0.36 N x 0.20 m = 0.1 J

The data should support the conclusion that the amount of work remains constant as the weight is increased.

19. Inconsistencies in the release of the tube probably account for most of the variability in the results. Other factors such as imprecise estimates of the height that the tube reaches, balls that do not weigh exactly as much as the tube, and failure to push the tube all the way to the maximum-force position are also possible sources of experimental error.

MODULE 3

Energy and Energy Conversions

◆Introduction

• When you drop a rubber ball, each bounce is lower than the one before it. Why?

• Why do you rub your hands together to keep them warm?

• Why does a billiard ball begin moving after it is struck by a cue ball?

The universe is composed of a combination of matter and energy. Matter is the "substance" of the universe, and energy is the "mover" of that substance.

Matter is easy to identify; it is tangible: We see, touch, smell, and taste matter constantly. Energy, on the other hand, is an abstract concept. It is when energy is being converted from one form to another that we are aware of it.

Since energy is most readily detectable when it is being transformed, we commonly use a definition for energy based on the physical concept of work:

Energy is the ability to do work.

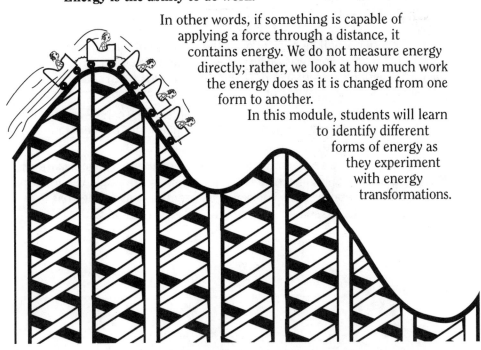

In other words, if something is capable of applying a force through a distance, it contains energy. We do not measure energy directly; rather, we look at how much work the energy does as it is changed from one form to another.

In this module, students will learn to identify different forms of energy as they experiment with energy transformations.

What are the different types of energy?

We are surrounded by many energy sources, but we usually notice the presence of energy only when it is converted from one form to another. The definition *energy is the ability to do work* does not give the entire energy picture. Scientists need a way to describe the energy content of the many *dissimilar* energy sources in the universe.

Rocks on top of a mountain, sunlight, beasts of burden, and lumps of uranium are very different entities, but each of them contains energy. All of them are capable of applying a force through a distance (doing work). The *energy* contained in rocks, sunlight, beasts, uranium, or any other object can be classified as being either:

kinetic energy, the energy due to an object's motion,
or
potential energy, the energy stored in both moving and non-moving objects due to an object's position or the arrangement of its parts.

Kinetic energy can be converted to potential energy and vice-versa. For example, the potential energy of a rock on a mountain changes to kinetic energy as the rock rolls down a slope. The rock can regain its original amount of potential energy if the kinetic energy of a moving horse is used to drag it back to the top of the mountain.

Sometimes the categories of kinetic and potential energy are too broad to be useful for describing individual cases of energy transformations. Analyzing energy transformations is easier when you know *specific forms* that energy can take. The following expanded definitions of kinetic and potential energy name some (but by no means all) of the forms that energy comes in.

Kinetic energy is the energy due to an object's motion.

Thermal energy,
light,
sound, and
electricity

are all forms of kinetic energy.

Potential energy is the energy due to an object's position or the arrangement of its parts.

Gravitational potential energy,
elastic potential energy, and
chemical energy

are all forms of potential energy.

When one of these forms of energy is used to do work, the same amount of energy will be present after the work is completed that was present before the work was begun. However, after the work is completed, the energy will be present in a *different form*. This is often stated as the conservation of energy principle:

Energy cannot be created or destroyed. Energy can be changed from one form to another, but the total amount of energy never changes.

This principle is a cornerstone for the study of energy transformations. *Proving* that energy is conserved is technically very difficult, even for very simple experimental systems. However, its usefulness in science has been seen again and again, and for now we will take its truth on faith.

◆Instructional Objectives

After completing the activities and readings for Module 3, students should be able to

• distinguish between kinetic energy and potential energy [Activities 9 and 10]

• identify different forms of kinetic and potential energy [Activities 11 and 12]

• demonstrate how energy can be changed from one form to another [Activities 11, 12, and 13]

◆Preparation

Study the following readings for Module 3:

Reading 6: Forms of Energy

Reading 7: Conserving Mass and Energy

Reading 8: Nuclear Energy: An Exception to the Conservation of Mass and Energy?

◆Activities

This module includes the following activities:

Activity 9: *Eureka!* #9—Kinetic Energy

Activity 10: *Eureka!* #10—Potential Energy

Activity 11: Demonstrating Energy Transformations

Activity 12: That's the Way the Ball Bounces

Activity 13: Does Heating Change a Ball's Energy?

ACTIVITY 9: VIDEOTAPE

Eureka! #9—Kinetic Energy

◆Background

This segment of *Eureka!* introduces the concept of **kinetic energy** by showing that moving objects have the ability to do work (apply a force through a distance). The term *energy* is derived from Greek words meaning *work in* (or *containing work*). Any object capable of performing work contains energy.

◆Time management

The running time of the videotape is five minutes. At least 15 minutes should be allotted to introduce, run, and discuss the videotape. You may wish to play the videotape at the end of a lesson to reinforce the concepts presented.

◆Comments on the videotape

The characters in the videotape demonstrate that energy can be converted to work and work can be converted to energy. The same unit, the joule (J), is used to measure both work and energy. Energy is not measured directly; instead, the energy content of a moving object (such as a billiard ball) is determined by measuring how much work it does.

Concept summary

- "Things which are moving have the ability to do work and to apply a force through a distance."*
- "They can be said to have 'work in' them."*
- "Another word for 'work in' is *energy*."*
- "Since energy is simply the ability to do work, and since work is measured in *joules*, energy is also measured in joules."*
- "The energy of movement is called *kinetic energy*."*

Eureka! Produced by TVOntario ©1981.

ACTIVITY 10: VIDEOTAPE

Eureka! #10—Potential Energy

◆Background

Segment 10 of *Eureka!* enlarges the concept of energy by showing that nonmoving objects (such as a boulder on a cliff) also have the ability to do work (apply a force through a distance). The type of energy possessed by nonmoving objects is **potential energy**.

◆Time management

The running time of the videotape is five minutes. At least 15 minutes should be allotted to introduce, run, and discuss the videotape. You may wish to play the videotape at the end of a lesson to reinforce the concepts presented.

◆Comments on the videotape

As in the previous segment of *Eureka!*, the characters demonstrate energy conversions with some clear and entertaining examples. For example, we see that kinetic energy is converted to work when a stone flung by a sling strikes a giant, bashing his nose. In addition, the characters show that work (flinging a stone straight up) can be converted to another form of energy, a form of stored energy: *potential energy*.

But potential energy does not remain in storage forever; the potential energy (contained by a stone in a high place) is converted to kinetic energy (when the stone falls). The kinetic energy contained by the rapidly falling stone can do work (bashing the little guy's nose).

The potential energy of the small stone thrown high into the air increases as its kinetic energy decreases. Potential energy resulting from an object's position in a gravitational field is classified as **gravitational potential energy**. The higher the object is above the surface of the Earth, the greater its gravitational potential energy will be.

In everyday life, we frequently encounter two other types of potential energy: **chemical energy**, energy which is stored in the chemical bonds that hold molecules together, and **elastic potential energy** that can be stored in materials such as rubber or steel. For example, when a spring is stretched, it stores elastic potential energy; when it snaps back to its original form, it performs work.

This videotape also illustrates and briefly discusses a very important idea:

> Energy can be converted from one form to another, but it is neither created nor destroyed.

This is another way of stating the conservation of energy principle.

Concept summary

• "Things which are not moving can have the ability to do work and to apply a force through a distance."*

• "The energy contained in things that are not moving is called *potential energy*."*

• "Potential energy can be converted to kinetic energy; kinetic energy can be converted to potential energy."*

Eureka! Produced by TVOntario ©1981.

ACTIVITY 11 WORKSHEET

Demonstrating Energy Transformations

◆Background

We may not be able to see **energy**, but we see the effects of energy transformations all the time. When a football player kicks a field goal, when an apple falls, and when you apply your brakes to slow a bike, energy is being transformed from one type to another. Our definition of energy is based on the physical concept of work:

Energy is the ability to do work.

If something is capable of applying a force through a distance (the definition of work), then it contains energy. In this series of activities you will investigate some simple cases of energy transformations.

◆Objective

To demonstrate some common energy transformations and learn ways to identify them

◆Procedure

Materials

Each group will need
• 2 ball bearings (1/2" in diameter)
• plastic tubing (at least 1 m in length; 3/4" inside diameter)

Vocabulary

• **Energy:** The ability to do work. Something contains energy if it can apply a force through a distance.
• **Kinetic energy:** The energy of moving objects.

Part I   Kinetic energy

Hold the plastic tubing in a U-shape. Drop a ball bearing into the tubing and let it come to rest at the bottom of the tube. Drop another ball bearing into the tube.

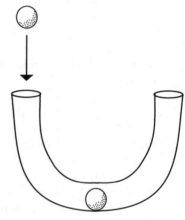

1. An object contains energy if it has the ability to do work. What does the moving ball bearing do that convinces you it has energy?

_____

_____

2. The energy of moving objects is called **kinetic energy**. Two factors determine the amount of kinetic energy an object possesses: how fast it is moving (its velocity), and the mass of the object. If a bowling ball and a basketball are moving at the same speed, the more massive bowling ball will have more kinetic energy. When the moving ball hits the stationary ball at the bottom of the tube, there is a transfer of energy. The moving ball has what kind of energy? The stationary ball acquires what form of energy?

_____

_____

3. What energy transformation is taking place in this demonstration?

_____

Part II   Work and heat energy

A. Hold your hands firmly together and slide your palms back and forth against each other.

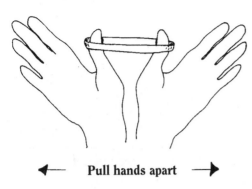

◄— Pull hands apart —►

B. Loop the rubber band loosely over your thumbs. Touch the rubber band to your lips. Quickly pull your thumbs apart, stretching the rubber band, several times. Touch the rubber band to your lips again.

C. Straighten the wire of the paper clip. Touch the wire to your lips. Bend the wire back and forth several times and touch it to your lips again.

4. When objects are bent or stretched or rubbed, work is usually being done on them, and they acquire energy. Some of that energy is in what form?

_____

_____

5. When a driver applies the brakes to stop a moving automobile, the brake pad rubs against a metal plate. Would you expect the brake pads to become hotter, colder, or stay about the same temperature while the car is slowing down?

_____

_____

Part III   Gravitational potential energy

A. Place your hand flat on a table with your palm up. Hold the apple 1 m above your hand and drop it. Catch the apple (without lifting your hand from the table, if possible).

B. Place your hand palm up on the table as before. Hold the apple 5 cm (0.05 m) above your hand and drop it. Catch it as before.

C. Place your hand palm up on the table as before. Hold the brick 5 cm (0.05 m) above your hand and drop it. Catch the brick (without lifting your hand from the table, if possible). *Do not drop the brick from a greater height!*

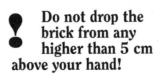 **Do not drop the brick from any higher than 5 cm above your hand!**

6. Nonmoving objects can also have the ability to do work: They can have **potential energy**. Potential energy that is due to an object's position in a gravity field (such as the Earth's) is called **gravitational potential energy**. The apple and the brick, when held stationary over the head, have

gravitational potential energy. That potential energy is transformed into what form of energy as the objects fall?

_____

_____

_____

_____

7. Catching an apple that falls 1 m is easy, but your hand would be injured if you dropped a *brick* on it from a height of 1 m. (*Do not try this!*) Which object, the apple or the brick, would transfer more energy to your hand? Which object, therefore, has more potential energy when held 1 m above your hand?

_____

_____

_____

Materials

Each group will need
• a spring-driven "jump-up" toy (or a "wind-up" toy)

Vocabulary

• **Elastic potential energy:** Potential energy that is stored in the molecules of objects that are stretched or compressed.

Part IV   Elastic potential energy

Place the toy on the table and press down until the spring is fully compressed. Move back and observe the motions of the toy.

8. Another type of stored energy found in nonmoving objects is **elastic potential energy**. This energy is stored in the molecules of objects that are stretched or compressed. The compressed spring of the toy has elastic potential energy. This potential energy is transformed into what form of energy the instant the toy leaves the table top? When the toy is at the top of its flight and is momentarily motionless, what form of energy does it have? If when the toy landed its spring was compressed, what form of energy would be stored in the spring?

_____

_____

_____

_____

Part V Chemical potential energy

Place baking soda in the bottle until the entire bottom of the bottle is covered with a layer of powder 1 cm to 2 cm deep.

Quickly pour about 100 ml of vinegar into the bottle and place the balloon over the mouth of the bottle. The balloon will inflate. (Hint: Use a funnel to transfer the vinegar quickly.)

9. Work was done on the balloon to stretch it out. Therefore, the balloon required elastic potential energy. That energy must have come from the chemicals that reacted. The energy stored in the chemicals is called **chemical potential energy**. Some of the chemical potential energy was transformed into what form of energy in the stretched balloon?

_____

_____

_____

_____

10. If you stick a pin into the inflated balloon, the elastic potential energy is transformed into what form(s) of energy?

_____

_____

_____

_____

Materials

Each group will need
• a small, empty soft drink bottle (16 oz or smaller)
• a balloon
• baking soda (sodium bicarbonate)
• 100 ml of vinegar
• **optional:** a small funnel

Vocabulary

• **Chemical potential energy:** Potential energy that is stored in the chemical bonds that hold an object's molecules together.

GUIDE TO ACTIVITY 11

Demonstrating Energy Transformations

◆What is happening?

The categories of energy presented in this activity, though not complete or comprehensive, are helpful when presenting the concept of energy transformation to students. Usually, students can easily observe changes in the position or temperature of an object, as well as the production of sound or light; by learning to associate these observations with categories of energy, students develop a way to identify energy transformations.

Once students are confident of their ability to identify energy transformations that have already occurred, they will be much more comfortable analyzing the forms of potential energy waiting to be transformed.

Part I of the activity never states that the kinetic energy of an object is one half the object's mass multiplied by the square of its velocity, but by emphasizing that it is the motion of the mass of the ball bearing that gives it the ability to do work—that gives it energy—students gain an intuitive feel for the fact that it is these two factors, velocity and mass, that determine kinetic energy.

Part II offers no explicit description of how increasing an object's kinetic energy increases the internal motion of the particles that make up the object, but students can observe the change in temperature directly and thereby conclude that kinetic energy is being transformed to heat the object.

Note: Heat *and* temperature *do not mean the same thing. The differences are discussed in Reading 6.*

Part III does not give the formula for gravitational potential energy (mgh), but the demonstration shows the importance of both mass and height, and that increasing either increases the energy. The apple weighs about 1 N; when it falls 1 m, it transforms about 1 J (1 N x 1 m) of gravitational potential energy into kinetic energy. The brick weighs about 20 N. After falling 0.05 m, it also transforms about 1 J of gravitational potential energy (20 N x 0.05 m) to kinetic energy. The impact of the brick should feel about the same as the impact of the apple.

Part IV introduces elastic potential energy to students. **Deformation** is the process of changing an object's shape by applying a force to it. Resistance to deformation is a property of all solids. Materials that are brittle simply break if sufficient force is applied.

In Part V, vinegar and baking soda undergo a chemical reaction, producing carbon dioxide gas. Another way to make use of chemical potential energy is by burning a substance. For example, the chemical bonds holding together the molecules in gasoline contain a large amount of chemical potential energy. Burning gasoline inside an engine releases the energy of these bonds and heats the air inside the cylinder. The kinetic energy of this heated air moves the pistons of the engine.

Materials that can store elastic potential energy all have internal molecular structures that resist being pushed together or pulled apart, but allow them to snap back to their original shape if the deforming force is removed.

Some types of energy are more difficult to detect than others. For example, measuring electrical energy often requires the use of some form of metering device. However, electrical energy's ability to do work can be demonstrated by using a comb charged with static electricity to pick up small bits of paper.

Other types of energy, such as nuclear energy, were not included in the types of kinetic or potential energy. Many classes can profit greatly by studying the production and uses of nuclear energy. However, discussing nuclear energy is easier when students are already familiar with the processes of energy transformation presented in this module.

◆Time management

One class period (40–60 minutes) should be enough time to complete the activity and discuss the results.

◆Preparation

This activity provides students with a means for using their observations to identify energy transformations. It is important to emphasize that energy is never lost or destroyed, only converted. As your students perform this activity, remind them that they are looking to see what happens to the the energy involved. To save time, you may wish to perform each part of the activity as a teacher demonstration and have the students follow along with their worksheets.

◆Suggestions for further study

The possible variations on the theme of demonstrating energy transformations are almost limitless. Encourage your students to bring in other toys or devices that transform energy and demonstrate them for the class. Many children's toys are entertaining because they convert energy in unique ways. Try to get your students to think about the energy transformations occurring all around them—energy should not just be something that they study during science class.

◆Answers

1. Since the moving ball bearing must have exerted a force through a short distance on the stationary ball bearing to get it moving, the moving ball bearing did work, and therefore had energy.

2. The moving ball has kinetic energy. The stationary ball acquires kinetic energy and some heat energy.

3. The primary transformation is the transfer of part of the kinetic energy of the moving ball to the ball at rest.

4. Some of the energy acquired from the work done by rubbing your hands together or stretching a rubber band, or bending a paper clip is in the form of heat energy.

5. The kinetic energy of the car is transferred as heat energy to the brakes—resulting in the slowing of the car and the increased temperature of the brakes.

6. The gravitational potential energy of the objects is converted to kinetic energy as the objects fall. This is the energy transformation taking place.

7. Near the surface of the Earth, two factors determine the amount of gravitational potential energy that is stored in an object: the *mass* of the object, and the *height* of the object above the Earth. Because the brick has more mass than the apple, it has more potential energy. Therefore, the brick will transfer more energy to your hand.

8. The spring stores elastic potential energy. When the toy is set in motion, much of the elastic potential energy is converted to kinetic energy. When the toy is at the top of its flight, it has the maximum gravitational potential energy. Just before the toy hits the table top, the gravitational potential energy is converted into kinetic energy. When the toy lands and its spring compresses, energy is stored in the form of elastic potential energy.

9. Some of the chemical potential energy of the chemicals was transformed into elastic potential energy when the balloon was stretched. Chemical potential energy is also transformed when you eat food. The chemical potential energy of the food is transformed into kinetic energy by your body, and this gives you the energy to lift objects.

10. Sticking a pin into the inflated balloon causes the elastic potential energy stored in the stretched balloon to be transformed into sound energy and kinetic energy. The balloon makes a popping noise (sound energy being released), and the pieces of the balloon move apart (releasing kinetic energy).

ACTIVITY 12 WORKSHEET

That's the Way the Ball Bounces

◆Background

Can you explain the bounce of a ball? When you hold a ball above the floor, its mass and height give it gravitational potential energy. When you drop it, an energy transformation takes place. What happens to the potential energy? Since the ball begins to move, some of the potential energy must become kinetic energy (the energy of moving objects). When the ball bounces back up and hangs motionless for an instant before falling again, it no longer has any kinetic energy. Has it all been transformed back into potential energy? If not, where did the energy go?

Materials

Each group will need
• a Superball™ or similar high-bouncing ball
• a squash ball
• a meter stick

◆Objective

To investigate the energy conversion of a bouncing ball

◆Procedure

1. Hold the meter stick vertically on a hard, flat surface. (A tile or concrete floor works best; using a table top is acceptable.)

Place the stick so that the 1-cm mark is nearest the floor and the 99-cm mark is nearest the ceiling.

2. Hold one of the balls 1 m (100 cm) above the surface, and about 15 cm in front of the meter stick. This is the *starting height*.

3. Release the ball. Avoid spinning the ball as you release it, or it will not bounce straight up.

Determine (in centimeters) how high the ball bounces. This is the *bounce height*.

4. Record the bounce height in the data table below:

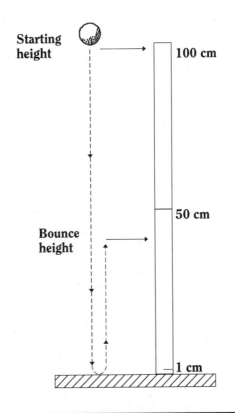

Data table

Bounce heights

Trial #	Starting height	Superball™ Bounce height	Squash ball Bounce height
1	100 cm	_____ cm	_____ cm
2	100 cm	_____ cm	_____ cm
3	100 cm	_____ cm	_____ cm
4	100 cm	_____ cm	_____ cm

5. Repeat steps 2, 3, and 4 three more times. Enter the bounce height for each of the four trials in the data table.

6. Determine the bounce height of the other ball. Repeat the procedure given in steps 2, 3, and 4. Record the results of your four trials of the second ball in the data table.

7. What is the average bounce height for a Superball™ dropped from a height of 100 cm?

 Average bounce height = _____ cm

8. What is the average bounce height for a squash ball dropped from a height of 100 cm?

 Average bounce height = _____ cm

9. Is the average bounce height for each of the balls greater than, less than, or equal to the original starting height of 100 cm?

_____

10. The higher the ball, the more gravitational potential energy the ball has stored in it.

When the ball reaches its average bounce height, does it have:

• the same amount of gravitational potential energy

• less gravitational potential energy

• more gravitational potential energy

than it had while it was being held 1 m above the surface?

_____

11. Since kinetic energy is determined by the mass and the velocity of an object, the faster the ball is moving, the greater its kinetic energy will be (as either mass or velocity increases, so does the kinetic energy).

When will the ball have the greatest amount of kinetic energy:

• before it is released,

• immediately after it is released,

or

• just before it strikes the surface?

_____

12. Complete the following statement: When the ball's gravitational potential energy is large, its kinetic energy will be _____ (*large, small*). When the ball is moving fast and it contains a large amount of kinetic energy, the amount of gravitational potential energy it contains will be _____ (*large, small*).

13. The kinetic energy of a moving ball is changed (transformed) into other forms of energy when it hits a surface and bounces.

Which of the following energy forms do you think is (are) produced as the ball strikes the floor?

_____ Sound	_____ Chemical potential energy
_____ Light	_____ Kinetic energy
_____ Electrical energy	_____ Nuclear energy
_____ Heat	_____ Elastic potential energy

Explain your choice(s).

_____

_____

_____

_____

_____

_____

_____

_____

_____

_____

GUIDE TO ACTIVITY 12

That's the Way the Ball Bounces

◆What is happening?

When a ball falls toward the floor, its *gravitational potential energy* is converted to *kinetic energy*. When it hits the floor, most of its kinetic energy is converted to *elastic potential energy*. Some kinetic energy is converted to sound (the thump that you hear); some of the kinetic energy heats the ball, the surrounding air, and the surface that the ball strikes. Some of its kinetic energy may enable the ball to move the surface that it lands upon. (This is why balls that are dropped on a flexible surface such as a cardboard box do not bounce very high—the ball's energy does work by displacing the box, leaving less energy to be converted to elastic potential energy.)

When the ball rebounds from the floor, the elastic potential energy that the ball stored on impact is converted back into kinetic energy and gravitational potential energy. A ball that is dropped (not *thrown* downward) *never* rebounds up to its original starting height. It will *rise less* on each bounce, because each time that it strikes the floor, some of the ball's kinetic energy is converted into forms that are not readily reconverted to the kinetic energy with which the ball rises again. The more efficiently a ball converts elastic potential energy back into kinetic energy (which is converted into gravitational potential energy), the higher it will bounce.

The balls used for this activity look very similar, but Superballs™ and squash balls bounce different distances when dropped from the same height. What can account for the different bounces?

Superballs™ bounce to about 90% of their original starting height when they are dropped. They bounce very high because they store elastic potential energy very efficiently and convert most of this elastic potential energy back into kinetic energy when they rebound from the floor. Superballs™ are solid balls that are made from a special type of rubber. They feel very hard when you squeeze them, and they deform (change shape) very little on impact.

Squash balls are very inefficient bouncers in comparison to Superballs™. A standard squash ball is hollow, and feels very soft when you squeeze it. When it strikes a surface, the air inside is compressed and heated, the ball's rubber molecules stretch, and it flattens out.

The energy gone toward deformation and heating is unrecoverable for the purpose of propelling the ball back toward its starting point; this energy is not reconverted into the kinetic energy with which the ball rises. Squash balls may only bounce to 15% of their original starting height.

◆Time management

One class period (40–60 minutes) should be enough time to complete the activity and discuss the results.

◆Preparation

The procedures for this activity were developed using Superballs™, produced by Wham-O™ Manufacturing Company and available at many toy stores, and standard SRA squash balls, manufactured by Slazenger.™ (Sporting goods stores that carry Slazenger™ tennis equipment can order the squash balls for you if they do not keep them in stock.) These balls are especially good for the activity because they are similar in size and appearance, but bounce very differently. However, the activity can be performed using *any combination of balls*.

A few suggestions for balls that your students may wish to test: golf balls, tennis balls (try using a new one and an old, "dead" tennis ball), balls from jacks sets, and high-bounce balls similar to Superballs™.

◆Suggestions for further study

In many sports, balls are struck by a club, bat, or part of a player's body. For these sports, the way that the ball bounces off the club greatly affects the play of the game. For example, only very powerful men are able to hit a baseball 450 ft for a homerun, but almost every amateur golfer from age 8 to 80 can hit a golf ball farther than that. How would golf change if golf balls were as "dead" as baseballs? Where would outfielders (and pitchers) have to stand if baseballs were as "lively" as golf balls?

Compare the bounce efficiency of the balls used in different sports such as golf, tennis, squash, baseball, softball, Whiffleball,™ racquetball, croquet, field hockey, ping pong, polo, lacrosse, and cricket (to name a few possibilities). Use a procedure similar to the one outlined in this activity to test how "live" or "dead" the balls used in these sports are. The bounce height divided by the starting height will give you a bounce coefficient for comparing the balls.

◆Answers

7. Average bounce height for Superball™: 91 cm.

8. Average bounce height for squash ball: 12 cm.

9. In every trial, the ball bounced to a height less than the original starting height of 1 m. The correct answer is *less than*.

10. The gravitational potential energy of an object is determined by the *mass* of the object and the *height* of the object above the Earth. The mass of the ball remains constant for this activity. The ball *never* bounces as high as the starting height of 1 m (unless it is *thrown* downward, which is an entirely different case). Since the average bounce height is always lower than the 1-m starting position, the ball must have less gravitational potential energy after it bounces. The correct answer is *less gravitational potential energy*.

11. The kinetic energy of the ball will be greatest when its mass and velocity are greatest, since kinetic energy is directly proportional to mass and velocity. The mass of the ball does not change while it is falling; therefore, it will have the greatest amount of kinetic energy when it reaches its greatest velocity. This occurs just before the ball hits the floor. (The ball accelerates as it falls because of the acceleration due to gravity. Therefore its velocity increases as it falls, reaching its highest value just before its motion is stopped.) The correct answer is *just before it strikes the surface*.

12. When the ball's gravitational potential energy is large, its kinetic energy will be *small*. When the ball is moving fast and it contains a large amount of kinetic energy, the amount of gravitational potential energy it contains will be *small*.

13. When it strikes the floor, the kinetic energy of the falling ball is transformed into *elastic potential energy* (stored in the stretchy rubber molecules that make up the ball, elastic energy "powers" the upward bounce) and *sound energy* (the thump you hear). Striking the surface also warms the ball (*heat energy*). You will probably not sense this change in temperature unless you bounce the ball rapidly and repeatedly, or you attach a sensitive thermometer to the ball. Some of the ball's *kinetic energy* is transferred to the surface that the ball strikes. If the surface is flexible, it will move when the ball strikes it; the impact will also warm the surface.

ACTIVITY 13 WORKSHEET

Does Heating Change a Ball's Energy?

◆Background

How can you make a dropped ball bounce higher other than by *throwing* it down? Throwing it down increases its kinetic energy (the energy of moving objects). Is there another way to increase the ball's energy so that it has more energy with which to bounce back up?

◆Objective

To increase a ball's energy in order to achieve a higher bounce

◆Procedure

1. Heat a container of water to about 60° C. Use the laboratory thermometer to check the temperature. Place the ball in this hot water bath while you set up the rest of the equipment. Use the tongs to turn the ball so the ball will become uniformly hot.

2. Hold the meter stick vertically on a hard, flat surface. Place the stick so that the 1-cm mark is nearest the floor, and the 99-cm mark is nearest the ceiling.

3. After the ball has been in the hot water for about *5 minutes*, use the tongs to remove it from the water.

Quickly dry the ball. Hold the ball 1 m (100 cm) above the surface, and about 15 cm in front of the meter stick. This is the *starting height*.

4. Release the ball. Avoid spinning the ball as you release it, or it will not bounce straight up.

Determine (in centimeters) how high the ball bounces. This is the *hot ball bounce height*.

5. Record the bounce height in the data table on the next page.

Starting height — 100 cm

50 cm

Bounce height

1 cm

Materials

Each group will need
• a squash ball (a Slazenger™ standard ball is recommended)
• a meter stick
• a container of ice water (mostly ice with a little liquid)
• a container of hot water (temperature should be about 50°–60° C)
• a thermometer (a laboratory thermometer calibrated to 100°C)
• tongs
• paper towels

❗ **60°C water can cause burns. Hot coffee is served at a temperature of about 60°C. Always use the tongs when moving the heated ball!**

Data table

Bounce height

Trial #	Hot ball bounce height	Cold ball bounce height
1	_____ cm	_____ cm
2	_____ cm	_____ cm
3	_____ cm	_____ cm
4	_____ cm	_____ cm

6. *Quickly* repeat steps 2, 3, and 4 three more times. Enter the bounce height for each of the four trials in the data table. If the ball begins to feel cool before you are able to complete all four trials, re-warm it in the hot water, then complete the four trials.

7. Place the ball in the ice water for *5 minutes*. Turn it so that all sides become equally cool.

8. Remove the ball from the ice and quickly dry it.

Drop the cold ball from a height of 1 m and measure its bounce height in the same way that you measured the bounce height of the hot ball.

Measure the bounce height of the cold ball a total of four times. Enter each bounce height in the data table. If the ball begins to feel warm between drops, place it back in the ice before completing your four trials.

9. What is the average bounce height for the hot ball dropped from a height of 100 cm?

Hot ball average bounce height = _____cm.

10. What is the average bounce height for the cold ball dropped from a height of 100 cm?

Cold ball average bounce height = _____cm.

11. Convert the average bounce heights from centimeters to meters by dividing the average height by 100. (Example: average bounce height = 32 cm ÷ 100 = 0.32 m)

Hot ball average = _____ cm ÷ 100 = 0._____ m.

Cold ball average = _____ cm ÷ 100 = 0._____ m.

12. Use the following formula to calculate the average amount of gravitational potential energy (GPE) of the ball at the high point of its bounce when it was hot or cold:

GPE = ball's weight (newtons) x bounce height (meters).

Squash balls weigh about 0.2 N. Use the hot ball and cold ball average bounce heights (in meters) that you calculated in Step 11 to complete the following calculations:

$GPE_{[hot]}$ = 0.2 N x 0._____ $m_{[hot\ ball\ average]}$ = _____ N • m = _____ J

$GPE_{[cold]}$ = 0.2 N x 0._____ $m_{[cold\ ball\ average]}$ = _____ N • m = _____ J

(The metric unit for energy is the joule [J]. 1 J = 1 N • m.)

Sample calculation: A squash ball bounces 0.32 m. At the top of its bounce, its gravitational potential energy is:

GPE = 0.2 N x 0.32 m = 0.64 N • m = 0.64 J

13. Which ball regained *more* gravititational potential energy when it bounced: the hot ball or the cold ball?

_____

14. Complete the following summary statement by filling in the blanks with the terms that best describe the behavior of the ball:

The weight of the ball is a/an _____ (*upward, downward*) force that is measured in newtons. Lifting a ball above a surface requires the application of an upward force through a distance. A force moving through a distance is, by definition, _____ (*energy, work*). The higher a ball is lifted above a surface, the _____ (*more, less*) work is required to raise it to that position, and the _____ (*more, less*) gravitational potential energy it contains.

Heating the ball _____ (*increases, decreases*) the upward distance that it bounces. The higher bounce of the heated ball shows that _____ (*more, less*) work is done on the heated ball than on the cold ball. Energy is defined as the ability to do work. Therefore, heating must have added energy to the ball.

GUIDE TO ACTIVITY 13

Does Heating Change a Ball's Energy?

◆What is happening?

The question posed by this activity is "does heating change a ball's energy?" In other words, can heating change a ball's ability to do work?

The formula for work is Work = Force x distance. You observed that the hot ball bounces to a greater height than the cold ball. A higher bounce requires more work, since the weight of the ball (a downward force) is moving through a greater distance (the height of the bounce).

The only variable being manipulated for this activity is the temperature of the ball. Therefore, you can infer that heating the ball supplies the energy doing the additional work on the hot ball (bouncing the hot ball higher).

Let's examine why this is the case. A ball held above a surface possesses gravitational potential energy (GPE). The amount of gravitational potential energy is determined by two factors: the weight of the ball and its height above the surface. This relationship can be stated as the equation GPE = weight x height.

When a ball is dropped, its gravitational potential energy is converted into kinetic energy. An instant before the ball strikes the floor, practically all of its gravitational potential energy has been converted into kinetic energy (relative to the floor). As the ball pushes against the floor, this kinetic energy is converted into elastic potential energy that is stored in the ball.

When the ball rebounds from the floor, the process is reversed: The ball's elastic potential energy is converted back into kinetic energy and the ball begins to rise. As it rises (constantly being slowed by the force of gravity), its kinetic energy is converted into gravitational potential energy. However, each impact "wastes" some energy by producing sound, heating the ball, and heating the surface that it strikes. This energy is not destroyed nor does it disappear; it is wasted in the sense that, by being converted to sound and heat energy, it is not available as elastic potential energy to be converted back into kinetic energy. Because of this wasted energy, balls never bounce back to the height from which they are dropped.

When a cold squash ball strikes a surface, a significant part of its kinetic energy is wasted by heating the rubber of the ball and heating the air trapped inside. However, when hot water is used to pre-heat the squash ball before dropping it, the heated ball rises to a greater height than the unheated ball. Therefore, we can infer that more energy, *the ability to do work*, is present in the ball after it has been heated than was present in the same ball when it was cooler. The energy that the squash ball gains from being heated does the work that bounces the ball higher.

Heating the squash ball makes the air molecules trapped inside move much faster. In other words, *heating the ball increases the kinetic energy of the air molecules inside it*. These faster-moving air molecules exert more pressure on the internal surfaces of the ball. You can sense this increased internal pressure by squeezing the hot squash ball—it feels harder than an unheated ball.

When the heated ball hits the surface, some of the extra kinetic energy that air inside has gained from the hot water leads to an extra push upward. Thus, the answer to the question posed by the title of this activity is *yes*—heating increases the energy of the squash ball, since heating increases the amount of work that it can do.

◆Time management

One class period (40–60 minutes) should be enough time to complete the activity and discuss the results.

◆Preparation

Before class, heat a large container of water to about 60° or 65° C. Use a laboratory thermometer to check the temperature. You may wish to place all the squash balls that students will use in this hot water bath yourself, rather than having students transport cups of scalding water to their lab stations. Use tongs to turn the balls so they will become uniformly hot. Remember, 60° C water can cause burns. Hot coffee is served at a temperature of about 60° C. Be careful when removing the balls.

Styrofoam cups make excellent ice baths for the balls. If you prefer, use a large pot or a cooler to make up a single ice bath for all of the balls.

Using squash balls for this activity is strongly recommended. Other types of balls *may* perform satisfactorily, but some will not. You may find that for certain types of balls (other than squash balls), the distance bounced by the hot ball is almost identical to the distance bounced by the cold ball. Other types of balls may bounce too erratically for students to obtain clear-cut results. Many factors besides the temperature of the ball can affect the way a ball bounces. Some of these factors are: the type of rubber used in the ball, whether the ball is solid or hollow, and whether or not it contains different layers (such as the strings wound around the core of a baseball).

◆Suggestions for further study

A golf ball company is attempting to sell you a golf ball heater for use on the golf course. They claim that the golf ball heater will significantly increase the distance the ball will fly. What controlled experiment would you perform to see if their claim is warranted? Can you detect any difference in bounce heights from dropping cold and hot golf balls?

◆Answers

Sample data

These data were collected using a Slazenger™ standard squash ball. The results you obtain may vary. The following bounce heights *are for reference only. They are not the only acceptable and correct answers for this activity.*

Hot ball: bounces ranged from 39 to 48 cm.

Cold ball: bounces ranged from 5 to 9 cm.

9. Hot ball average bounce height = 46 cm.

10. Cold ball average bounce height = 6 cm.

11. Hot ball average bounce height in meters = 0.46 m.
    Cold ball average bounce height in meters = 0.06 m.

12. $GPE_{[hot]} = 0.2$ N x 0.46 $m_{[hot\ ball\ average]} = 0.92$ N•m $= 0.92$ J

    $GPE_{[cold]} = 0.2$ N x 0.06 $m_{[cold\ ball\ average]} = 0.12$ N•m $= 0.12$ J

13. The hot ball regained more gravitational potential energy.

14. The weight of the ball is a *downward* force that is measured in newtons. Lifting a ball above a surface requires the application of an upward force through a distance. A force moving through a distance is, by definition, *work*. The higher a ball is lifted above a surface, the *more* work is required to raise it to that position, and the *more* gravitational potential energy it contains.

Heating the ball *increases* the upward distance that it bounces. The higher bounce of the heated ball shows that *more* work is done (by the floor) on the heated ball than on the cold ball. Energy is defined as the ability to do work. Therefore, heating must have added energy to the ball.

MODULE 4

Machines Are Simple!

◆Introduction

• Machines actually *increase* the amount of work required to perform a task. Why then do we use machines?

• Scientists consider an inclined plane to be a machine, even though it does not have moving parts, flashing lights, or require electricity. What makes an inclined plane a machine?

• All machines waste some energy. Why would anyone want to use a machine?

Most people, when asked to name a machine, will mention a complex device like an automobile, a computer, or an air conditioner. The functioning of each of these complex machines depends on small, simple sub-units: simple machines.

Any device that helps you do work is a machine. Machines can accomplish this by

• transforming energy from one form to another,

• transferring energy from one place to another,

• multiplying force,

• changing the speed of an object, or

• changing the direction of a force.

In this module, you will investigate the properties of several simple machines and design and build a machine that can accomplish a complex task.

◆Instructional Objectives

After completing the activities and readings for Module 4, students should be able to

• describe how a simple machine can help you do work [Activities 14, 15, and 16]

• demonstrate that for a specific task, using a machine may decrease the effort required but actually *increase* the total amount of work required to accomplish the task [Activities 17 and 18]

• assemble a complex machine by combining several simple machines [Activity 19]

◆Preparation

Study the following readings for Module 4:

Reading 9: Scientific Literacy for All

Reading 10: Opening Up the Black Box of a Complex Machine

◆Activities

This module includes the following activities:

Activity 14: *Eureka!* #11—The Inclined Plane

Activity 15: *Eureka!* #12—The Lever

Activity 16: *Eureka!* #13—Mechanical Advantage and Friction

Activity 17: Just "Plane" Work

Activity 18: Testing a Paper Clip Pulley System

Activity 19: "Unsimple" Machines: Designing a "Rube Goldberg" Device

ACTIVITY 14: VIDEOTAPE

Eureka! #11—The Inclined Plane

◆Background

The term *machine* makes most people think of something like an automobile or a piece of industrial equipment. Scientists and engineers define machine differently:

> A machine is an object that *changes the force* required to do some task.

Eureka! #11 shows why the inclined plane is an example of the scientific definition of a machine.

◆Time management

The running time of the videotape is five minutes. At least 15 minutes should be allotted to introduce, run, and discuss the videotape. You may wish to play the videotape at the end of a lesson to reinforce the concepts presented.

◆Comments on the videotape

Confronted with a barrel that is too massive to lift directly into a truck, the videotape character explores two options: cut the barrel into four smaller pieces and lift each piece into the truck separately, or employ the inclined plane to reduce the force required to raise the barrel. Since cutting things apart is often impractical, he opts for the inclined plane.

Is our cargo handler getting something for nothing by using the inclined plane? Is the inclined plane somehow adding energy to the system to allow him to lift the barrel? The answer is *No!* to both of these questions.

Energy is the ability to do work. Work is equal to the force applied times the distance an object moves (W = F x d). The loader does 800 J of work when lifting the barrel directly into the truck:

> 800 N x 1 m = 800 J.

Using the inclined plane to lift the barrel *reduces the force* required to move the barrel to 200 N, but it *increases the distance* that the barrel moves to 4 m. Therefore, the work done is

> 200 N x 4 m = 800 J,

which is the *same amount* of work required to lift the barrel directly into the truck. So we see that the inclined plane *does not* add energy to the system.

In order to keep the discussion simple, the videotape does not mention the effects of friction. In fact, because of friction, using the inclined plane would require slightly *more* work than lifting the barrel directly onto the truck. Friction affects all machines. No machine is 100 percent efficient, so using any machine always increases the total work being done.

Why would anyone choose to use a machine that makes you do more work to complete a task? The answer to this question depends on the type of task being attempted. Many tasks such as changing a tire on a car would be impossible for one person to perform without using a machine. Raising a car wheel off of the pavement requires the application of more force than an unassisted person can produce. However, by using a jack, (which is a modified form of an inclined plane) lifting the car becomes fairly easy. Paying the price of doing a small amount of extra work to overcome the friction of the jack is worth it to a stranded motorist.

Concept summary

• "A machine is something that helps you use your energy more effectively."*

• "One of the simplest machines is the inclined plane, which allows you to trade increased distance for decreased effort or force."*

• "By doubling the distance you move a thing, you can half the force you need to exert."*

Eureka! Produced by TVOntario ©1981.

ACTIVITY 15: VIDEOTAPE

Eureka! #12—The Lever

◆Background

All of the machines that we use are derived from two basic types of simple machines: the inclined plane and the lever.

Levers come in a variety of sizes and shapes and range from bottle openers to baseball bats to drawbridges. No matter what their size or function, all levers have three basic parts:
• the **fulcrum** or pivot point,
• the **effort** arm, which is where the force is applied, and
• the **resistance** arm, which is where the resistance or load is located.

◆Time management

The running time of the videotape is five minutes. At least 15 minutes should be allotted to introduce, run, and discuss the videotape. You may wish to play the videotape at the end of a lesson to reinforce the concepts presented.

◆Comments on the videotape

This program uses a teeter-totter to illustrate some of the things that levers can do. A teeter-totter with arms of unequal length allows the small character to balance the large character, or even to launch him into the air. The relative lengths of the arms of the lever determine the force required to move a given load. For this type of lever, moving the fulcrum nearer to the resistance (load) *decreases* the amount of force that must be applied to the effort arm.

The film shows an example demonstrating the *principle of the lever*:

> The *longer* the arm of the lever to which a force is applied, the *less* that force need be.

For example, if the effort arm is *twice as long* as the resistance arm, a force of 100 N on the effort arm will lift a load of 200 N placed on the resistance arm. However, the effort must be applied through a distance of 1 m in order to move the resistance 0.5 m.

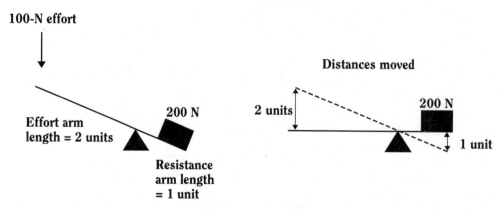

100-N effort

Effort arm length = 2 units

200 N

Resistance arm length = 1 unit

Distances moved

2 units

200 N

1 unit

ACTIVITY 16: VIDEOTAPE

Eureka! #13—Mechanical Advantage and Friction

◆Background

All machines are affected by friction to some extent. Since no machine is 100 percent efficient, a reasonable question for a person using a machine to ask is "How much easier does this machine make the task I'm performing?" Calculating the **mechanical advantage** of a machine provides a numerical answer to this question.

◆Time management

The running time of the videotape is five minutes. At least 15 minutes should be allotted to introduce, run, and discuss the videotape. You may wish to play the videotape at the end of a lesson to reinforce the concepts presented.

◆Comments on the videotape

In this segment of *Eureka!*, two professors use a spring scale to measure the force required to lift a book. They raise the book to the same height (10 cm) in three different ways:
• lifting it straight up,

• raising it using a lever, and

• sliding it up an inclined plane.

Lifting the book straight up does not require the use of a machine. (The spring scale is used to measure the force being exerted; it is not used to lift the book.) The work done ($W = F \times d$) is equal to the weight of the book (24 N) times the distance it moves (10 cm, or 0.1 m). Multiplying these gives:

$$\text{Work}_{[lifting]} = 24 \text{ N} \times 0.1 \text{ m} = 2.4 \text{ N} \cdot \text{m} = 2.4 \text{ J}.$$

There is friction between the air and the book, but this force is so small that it can be ignored. Lifting the book straight up is the most efficient means of moving it 10 cm, and for small objects like books, that is what we usually do.

A well-designed lever (such as the one the scientist uses to lift the book) has very little friction between the fulcrum and the arm. The work put into moving a lever is almost exactly equal to the work output. Using a lever is *almost* as efficient as lifting an object straight up. You can think of the work input as the amount of work done when using the machine, and work output as the amount of work that would be required for the same task *without* the use of the machine.

Why use a lever if its work output is about equal to the work input? The answer is that even though the work input and output are about equal, *much less force* (only 6 N) is required when using the lever to raise the book. The effort arm of the lever moves through a *greater distance* (0.4 m) to accomplish the same work that was done by lifting the book 0.1 m straight up:

$$\text{Work}_{[input]} = 6 \text{ N} \times 0.4 \text{ m} = 2.4 \text{ N} \cdot \text{m} = 2.4 \text{ J},$$
$$\text{Work}_{[output]} = 24 \text{ N} \times 0.1 \text{ m} = 2.4 \text{ N} \cdot \text{m} = 2.4 \text{ J}.$$

Calculating a machine's mechanical advantage is a shorthand way of describing the relationship between its output force and input force.

Concept summary

• "If you divide the force you get out of a machine by the force you put into it, you will discover the mechanical advantage of that machine."*

• "But since every machine involves contact between moving surfaces, its mechanical advantage will always be reduced to some extent by the force of friction."*

• "There's a lot of friction involved with the inclined plane, but hardly any with the lever."*

Eureka! Produced by TVOntario ©1981.

Mechanical advantage (M.A.) is calculated as follows:

$$\text{M.A.} = \frac{\text{Output force}}{\text{Input force}} .$$

The lever shown in the film has a M.A. of 4:

$$\text{M.A.} = \frac{\text{Output force}}{\text{Input force}} = \frac{24 \text{ N}}{6 \text{ N}} = 4.$$

Dragging the book up the inclined plane requires considerably more work than lifting it 10 cm straight up or raising it with a lever. Friction significantly increases the input force necessary to move the book. The reading from the spring scale can be used to calculate the inclined plane's mechanical advantage:

$$\text{M.A.} = \frac{\text{Output force}}{\text{Input force}} = \frac{24 \text{ N}}{8 \text{ N}} = 3.$$

Since the book is moved 0.4 m in order to raise it to a final height of 0.1 m, the work being done is:

$$\text{Work}_{\text{[input]}} = 8 \text{ N} \times 0.4 \text{ m} = 3.2 \text{ N} \cdot \text{m} = 3.2 \text{ J},$$
$$\text{Work}_{\text{[output]}} = 24 \text{ N} \times 0.1 \text{ m} = 2.4 \text{ N} \cdot \text{m} = 2.4 \text{ J},$$

Using the inclined plane increases the total work required to raise the book by 0.8 J. In other words, 0.8 J is the energy price for using the inclined plane. This does not seem like a good deal for moving a small object like a book, but if the object being lifted were a grand piano, the energy price would be a bargain.

ACTIVITY 17 WORKSHEET

Just "Plane" Work

◆Background

A **machine**, by its scientific definition, is not necessarily a complicated device like a car or a computer. In order for something to be considered a machine, all it has to do is change the force necessary to perform some task. For example, something as simple as an inclined plane is a machine. How can it change the force required to do something if it has no moving parts? And how does it affect the amount of *work* involved in a task? This activity will show you that an inclined plane increases the *work* necessary to move an object, but decreases the *force*.

◆Objective

To investigate how an inclined plane affects the work and the force required to move an object

◆Procedure

1. Choose one end of the board to be the Starting line. Tear or cut the masking tape into 6 strips, each about 1 cm long. Measure 15 cm from the starting line. Stick a piece of tape on the edge of the board to mark the 15 cm distance.

Materials

Each group will need

• a toy car, toy truck, or similar object that rolls and to which a spring scale can be attached. The car should weigh 3–4 N.

• a smooth board that is wide enough for the car to roll on. It should have a minimum length of 122 cm.

• a spring scale (calibrated to 5 N)

• a meter stick

• books for supporting the board

• 6 cm of masking tape

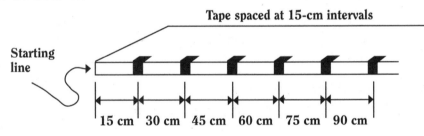

Tape spaced at 15-cm intervals

Starting line

15 cm 30 cm 45 cm 60 cm 75 cm 90 cm

Stick the remaining five strips of tape on the edge of the board at 15-cm intervals. When you finish, the 6 tape markers will be located 15, 30, 45, 60, 75, and 90 cm from the Starting line.

Vocabulary

• **Machine:** An object that changes the force necessary to perform some task.

2. Make a stack of books 20 cm high. Prop the board on this stack of books so that its 90-cm tape marker is directly above the point of support. Attach the spring scale to the car. Set the car on the board. Place the rear wheels of the car on the Starting line.

3. Pull on the spring scale, moving the car up the inclined plane at a steady speed. *While the car is moving at a steady speed,* read the force in newtons. Record this force reading in the data table.

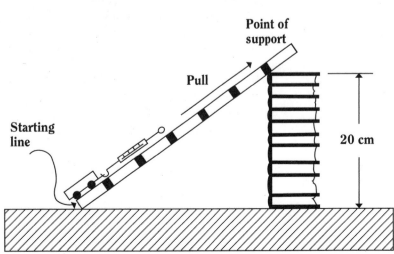

Point of support

Pull

Starting line

20 cm

Stop the car when its rear wheel reaches the 90-cm mark. When the car reaches this line, it will be 20 cm above its starting point.

Data table

Work required to raise a car 20 cm

Point of Support	Force required to move car		Distance the car moves		Work to reach height of 20 cm
90-cm mark	_____ N	x	0.9 m	=	_____ J
75-cm mark	_____ N	x	0.75 m	=	_____ J
60-cm mark	_____ N	x	0.6 m	=	_____ J
45-cm mark	_____ N	x	0.45 m	=	_____ J
30-cm mark	_____ N	x	0.3 m	=	_____ J

4. Move the inclined plane so that the 75-cm mark is above the support point on the pile of books. Place the car on the Starting line as you did before, and pull the car until its rear wheel reaches the 75-cm mark. Determine how much force is required to move the car up the inclined plane at a constant speed, and record this reading in the data table.

5. Move the inclined plane again, and make force readings with the board supported at the 60-, 45-, and 30-cm marks. Enter the force readings that you obtain in the data table.

6. For each of the settings of the inclined plane, calculate the amount of work (in joules) required to raise the car 20 cm. To do this, multiply the force reading you obtained by the distance that the car moved along the plane in the direction of the force.

Record the results of your calculations in the data table.

7. On the basis of your observations and calculations, complete the following statements:

Changing the point of support of the inclined plane changes the distance that the car must roll to reach a point 20 cm above its starting point. As the *distance* the car rolls *increases*, the amount of *force* required to raise the car _____ (*increases, decreases, stays about the same*).

As the point of support of the plane is changed from 90 cm down to 30 cm, the *work* required to raise the car 20 cm _____ (*increases, decreases, stays about the same*).

8. How much work would it require to *lift* the car 20 cm *without using the inclined plane*? To answer this question, simply attach the spring scale to the car and lift the car 20 cm high. Read the force indicated on the scale. The work performed is

Reading on the scale _____ N x 0.2 m = _____ J.

9. How does the work you did by lifting the car directly compare with the

work you did using the inclined plane?

_____

_____

10. Is the inclined plane the *only* machine that you are using to move the car for this experiment? Remember, a machine is defined as an object that *changes the force* required to do some task. What about the wheels on the car—are they machines? Test the effects that the wheels have on the force required to raise the car 20 cm with the following procedure.

Set the inclined plane on the 75-cm mark. Attach the spring scale to the car as you did before, but this time place the car *upside down* on the inclined plane, so that its wheels *do not roll* along the board. While the car is moving at a constant speed, read the force required to drag it to the 75-cm mark (a height of 20 cm). The work performed is

Reading on the scale _____ N x 0.75 m = _____ J.

11. Does *dragging* the car along upside down to a height of 20 cm require the same amount of work as *rolling* it a height of 20 cm? Why or why not?

_____

_____

_____

_____

12. An inclined plane is a simple machine. Should the *wheels* on the car also be classified as machines? Explain why or why not.

_____

_____

_____

_____

GUIDE TO ACTIVITY 17

Just "Plane" Work

◆What is happening?

If the board being used for this experiment is smooth and the car has low-friction wheels, the total amount of work required to raise the car 20 cm will remain fairly constant for all the settings of the inclined plane. The work done while using the inclined plane will be almost the same as the work required to lift the car directly to a height of 20 cm.

More force is required to pull a car up a steep incline (for example, when the board is set at the 30- or 45-cm marks) than is required to pull it up a shallower slope (such as the 90-cm setting). However, the *total* amount of work ($W = F \times d$) *remains about the same* because the *distance* that the car has to move *decreases* as the slope becomes steeper and the *force increases*.

The wheels of the car play a major role in determining how much force is required to pull the car along the inclined plane. The wheels reduce the friction between the car and the plane. Dragging the car upside down requires a much greater force than rolling it up the same slope. It takes considerably more work to drag the car than to roll it.

◆Time management

One class period (40–60 minutes) should be enough time to complete the activity and discuss the results.

◆Preparation

Any car used for this activity must
• have fairly low-friction wheels, so that it rolls smoothly.

• be light enough for its weight to register accurately on the spring scale even when the car is moving up a shallow inclined plane. (If you are using a 5-N scale, the car should weigh less than 5 N.)

• be heavy enough to deflect the pointer on the spring scale even when the car is moving up a shallow inclined plane. (If the car weighs only 1 N, very little force will be required to move it upward when the board is set on the 90-cm mark. The scale may not register this small force very accurately. There are two possible solutions to this problem: Tape weights such as lead fishing sinkers to the car, or use a more sensitive spring scale.)

While anything that meets the above requirements may be used in this activity, laboratory testing carts such as Hall's carriages are ideal. They are available through many scientific suppliers, but their cost may prove prohibitive. Plastic roller skates or toy cars with low friction wheels will also work very well.

◆Suggestions for further study

In this activity you have observed a number of forces needed to pull the car up the incline. Choose some middle value force which you have not observed and assume that this force is the maximum force that can be exerted to pull the car up the hill. (In a real situation this maximum force might be the maximum force which the engine of a car can produce or the maximum strength of a tow rope used to pull the car up the hill.) With this maximum force what is the steepest hill that can be climbed? Measure steepness in terms of angle.

Sometimes when we climb a hill it seems that it is harder to climb near the top of the hill. Consider what you have seen in this activity or make more observations to see if *harder* means *more force required*. In other words, on a hill of given steepness, is the force required to move near the top of the hill greater than the force required to move near the bottom of the hill? If not, why might it *seem* harder to climb near the top of the hill?

◆Answers

The following represent only *one possible set of observations*. Your results should follow the same general pattern illustrated here, but *the actual numerical results you obtain will vary*, depending on the type of car you use.

Work required to raise a car 20 cm

Point of support	Force required to move car		Distance the car moves		Work to reach height of 20 cm
90-cm mark	0.9 N	x	0.90 m	=	0.81 J
75-cm mark	1.1 N	x	0.75 m	=	0.83 J
60-cm mark	1.3 N	x	0.60 m	=	0.78 J
45-cm mark	1.7 N	x	0.45 m	=	0.77 J
30-cm mark	2.4 N	x	0.30 m	=	0.72 J

7. Changing the point of support of the inclined plane changes the distance that the car must roll to reach a point 20 cm above its starting point. As the distance the car rolls *increases*, the amount of *force* required to raise the car *decreases*. As the point of support of the plane is changed from 90 cm down to 30 cm, the work required to raise the car 20 cm *stays about the same*.

Note: You may notice from your data that the amount of *work* required actually *decreases* slightly as the slope of the board becomes steeper. This decrease reflects the decreased rolling resistance of the wheels at steeper angles. When the incline is steep, the spring scale supports some of the car's weight. Each wheel carries less weight and can therefore roll more easily, decreasing the total work required to move the car.

8. The car used to collect these data weighs *3.25 N*. Directly lifting it 20 cm requires

   3.25 N x 0.2 m = *0.65 N • m = 0.65* J of work.

9. Rolling the car up the incline plane to a height of 20 cm required between *0.72 J* and *0.81 J* of work. Directly lifting the car 20 cm requires less work (only *0.65 J*) than any setting of the incline plane.

10. The work required to drag the car upside down is

    0.9 N x 0.75 m = 0.68 N • m = 0.68 J.

11. Dragging the car up the ramp requires about twice as much work as rolling it. More work is required to overcome the increased force of friction acting on the car while it is being dragged up the ramp.

12. The wheels are machines since they fit the scientific definition of a machine: They reduce the force necessary to perform a task. By reducing friction between the car and the inclined plane, the wheels help to reduce the force required to move the car upward. (Note: You never get something for nothing when using machines. Since the wheels have mass, they also slightly increase the force required to move the car. However, the substantial decrease in friction makes using wheels worth it.)

ACTIVITY 18 WORKSHEET

Testing a Paper Clip Pulley System

◆Background

Have you ever wondered why a pulley allows a couple of people to lift a piano to the roof of a building when they would never even be able to lift the piano off the ground by themselves? The pulley doesn't make the person pulling it any stronger, the piano doesn't weigh any less, and the piano still has to be moved as far. What is different? If the people don't exert any more force when using the pulley than they would without it, but the piano moves when the pulley is used, then the pulley must somehow *decrease the force required to move the piano*. This means the pulley must be a *machine*.

◆Objective

To construct and test a paper clip pulley system

◆Procedure

1. Construct a resistance for testing the pulleys as follows: String the five lead weights on the 10-cm length of heavy twine. Tie a knot in the twine, forming a loop.

2. Weigh the resistance on the spring scale.

   Weight of resistance = _____ N.

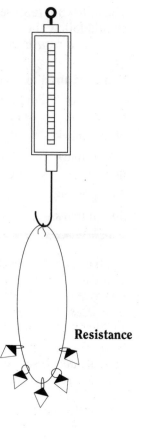

Resistance

Materials

Each group will need
• 5 lead weights (2-oz fishing sinkers)
• a 10-cm length of heavy twine
• masking tape
• 1 standard wire paper clip
• 1 small binder clip
• 1-m length of sewing thread
• a spring scale calibrated to 5 N
• a meter stick

Vocabulary

• **Mechanical advantage:** A measure of the advantage gained by using a machine to change the force necessary to perform a task. Mechanical advantage is calculated by dividing the force that the machine produces by the force applied to the machine.

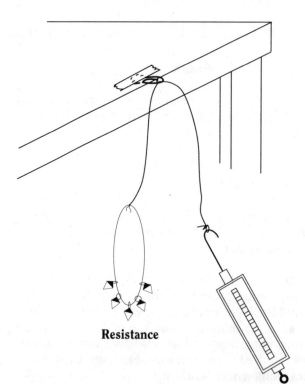

Resistance

3. Tape a paper clip to the edge of a table so that one end of the clip hangs off the table, forming an eye.

Run the sewing thread through the eye. Tie one end of the thread to the twine of the lead weight resistance, and attach the spring scale to the other end.

4. Pull the scale so that the resistance is lifted 20 cm off the floor.

How far does the scale have to move in order to lift the resistance 20 cm?

Scale moves _____ cm.

How many newtons of effort are required to lift the resistance?

Effort force = _____ N.

5. Calculate the **mechanical advantage** of this *single fixed pulley* system as follows:

Mechanical advantage = $\dfrac{\text{resistance}}{\text{effort.}}$

In this case, the resistance (abbreviated R) is equal to the weight (in newtons) of the fishing sinkers.

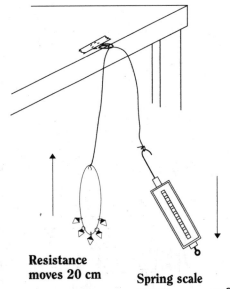

Resistance moves 20 cm

Spring scale moves _____ cm?

The effort force (abbreviated E) is the force reading on the scale while the object is being lifted. For the single paper clip,

Mechanical advantage = $\dfrac{R}{E}$ = $\dfrac{\_\_\_\_ \text{ N}}{\_\_\_\_ \text{ N}}$ = _____ .

Run thread through handle

Attach resistance to binder clip

6. Untie the resistance from the thread. Leave the spring scale attached to the other end of the thread.

As the diagram illustrates, attach the resistance weights to a binder clip by clamping the twine in the jaws of the clip.

Run the sewing thread through one of the handles of the binder clip.

Tape the loose end of the sewing thread to the table close to the paper clip.

7. Pull the scale so that the resistance is lifted 20 cm off the floor.

How far does the scale have to move in order to lift the resistance 20 cm?

Scale moves _____ cm.

How many newtons of effort are required to lift the resistance?

Effort force = _____ N.

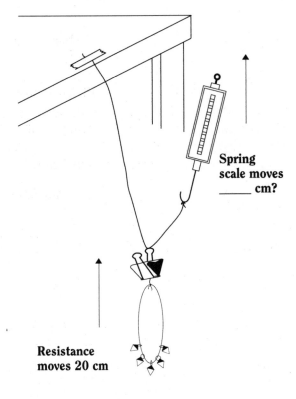

Spring scale moves _____ cm?

Resistance moves 20 cm

8. Calculate the mechanical advantage of this *single moveable pulley*:

Mechanical advantage = $\dfrac{R}{E}$ = $\dfrac{\underline{\quad} N}{\underline{\quad} N}$ = $\underline{\qquad}$.

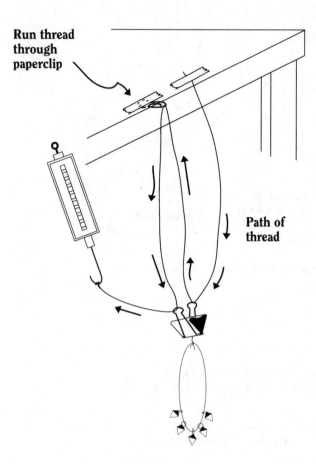

Run thread through paperclip

Path of thread

9. Remove the spring scale from the sewing thread. Leave the other end of the thread running through the handle of the binder clip and taped to the table.

Run the free end of the thread through the eye of the paper clip that you taped to the table in Step 3.

Now run the sewing thread through the second handle of the binder clip. Reattach the spring scale to the end of the thread.

10. The sewing thread supporting the resistance should now form a W-pattern. Be sure the thread is not twisted, and that the clip holding the resistance can slide freely.

11. Pull the scale so that the resistance is lifted 20 cm off the floor.

How far does the scale have to move in order to lift the resistance 20 cm?

Scale moves _____ cm.

How many newtons of effort are required to lift the resistance?

Effort force = _____ N.

12. Calculate the mechanical advantage of this *double moveable pulley*:

Mechanical advantage = $\dfrac{R}{E}$ = $\dfrac{\underline{\quad} N}{\underline{\quad} N}$ = $\underline{\qquad}$.

13. On the basis of your observations, complete the following statement:

If the *effort force* and the *resistance* move the same distance when the string is pulled, the pulley system will have a *mechanical advantage* _____ (*greater than, less than, equal to*) 1.

If the effort force moves a *greater distance* than the resistance moves, the pulley will have a mechanical advantage of _____ (*greater than, less than, equal to*) 1.

GUIDE TO ACTIVITY 18

Testing a Paper Clip Pulley System

◆What is happening?

Simple machines provide mechanical advantage by allowing the user to apply a *smaller* amount of *force* over a *greater* amount of *distance*. Pulley systems show the relationship between mechanical advantage and distance more clearly than other simple machines.

The paper clip and paper clamp were used in place of conventional pulleys for two reasons:
• to demonstrate that the wheel of a conventional pulley is not an essential part of the machine. The wheel does not do any work; its function is to reduce friction acting on the rope.

• to provide a means of studying pulleys that does not require the purchase of a large number of conventional pulleys. This allows more people to have hands-on experience with pulleys for a very modest expenditure for materials.

◆Time management

One class period (40–60 minutes) should be enough time to complete the activity and discuss the results.

◆Preparation

There are two sources of systematic error built into the procedure for this activity. One is easily cured, the other can be made small enough that it does not greatly affect the results.
• When determining how far the *scale* moves in comparison to the movement of the *resistance* (see steps 4, 7, and 11), the method of measurement can affect the results. The spring of the scale stretches (in some cases, several centimeters) before the resistance itself moves. This can be prevented by *first* pulling down on the scale until the resistance just starts to move, *then* measuring the motion of the scale and the resistance.

Alternatively, hold onto the *string* rather than the scale and measure how far your *fingers holding the string* move. You may wish to demonstrate these measuring techniques to your students.
• Most laboratory scales cannot be adjusted to measure force accurately while they are being held upside down. The first pulley set-up (see step 3) requires you to pull the scale *downward*. When the scale is held in this position, the reading on the scale is the effort force *minus* the weight of the hook and metal bar to which the spring is attached.

However, if you use a resistance that weighs about 250 g (as suggested), the scale reading will be incorrect by only about 10–20 percent. Mathematically correcting the scale reading may cause some students to become confused, and since the relative amount of error is small, we recommend living with this particular error.

◆Suggestions for further study

In this activity, mechanical advantage has been calculated by dividing the resistance force by the effort force. How might you use distances moved by the effort and resistance forces to calculate a mechanical advantage? For example, if the resistance force moves 20 cm and the effort force moves 80 cm (as in the double moveable pulley), what might be the mechanical advantage of the pulley? How does this mechanical advantage compare with the one you calculated from the forces acting in the double moveable pulley system? Why are the mechanical advantages different? Which is a better method of calculating mechanical advantage?

In spite of what you might see bionic creatures do on TV, they can only exert a two-handed downward push or pull which is equal to their weight. If one of these creatures weighs 200 pounds and lifts a 1,000-pound safe with a pulley system, what must be the minimum mechanical advantage of the pulley system?

◆Answers

Note: The following calculations are based on *one possible set of observations. They are not the only acceptable answers for this activity.* The actual data that you collect will differ from the results listed below.

2. Weight of resistance = *3.0* N (based on using five fishing sinkers, each weighing about 2 oz).

4. The scale moves *20* cm when the resistance moves 20 cm. The effort force required is *4.0* N.

5. Mechanical advantage = *3.0 N/4.0 N = 0.75* for a single *fixed* pulley.

7. The scale moves *40* cm when the resistance moves 20 cm. The effort force required is *1.9* N.

8. Mechanical advantage = *3.0 N/1.9 N = 1.6* for a single *moveable* pulley.

11. The scale moves *80* cm when the resistance moves 20 cm. The effort force required is *1.0* N.

12. Mechanical advantage = *3.0 N/1.0 N = 3* for a double *moveable* pulley.

13. If the effort force and the resistance move the same distance when the string is pulled, the pulley system will have a mechanical advantage of *less than* 1.

If the effort force moves a greater distance than the resistance moves, the pulley will have a mechanical advantage of *greater than* 1.

ACTIVITY 19 WORKSHEET

"Unsimple" Machines: Designing a "Rube Goldberg" Device

◆Background

Rube Goldberg's syndicated cartoons of the 1930s through the 1960s often featured a "Weekly Invention" consisting of an elaborate, humorous device for solving problems such as walking safely on icy pavements. Goldberg's "unsimple" machines became so widely known (and copied) that the term "Rube Goldberg" device is now used to describe any elaborate, inefficient, or awkward-looking machine. The U.S. Navy copied Goldberg's style when they included the following cartoon in a training manual.

Porthole Closer—Blanket Puller-Upper*

This machine was invented by a guy named Oscar. Sea water entering open port is caught in helmet (1) hung on rubber band. Rubber stretches and helmet is pushed down against shaft of Australian spear (2). Head of spear tips over box of bird seed (3) which falls in cage (4) where parrot (5) bends over to pick it up. Board strapped on parrot's back pulls on string (6) which releases arrow (7) and slams the port shut. Breeze from closing port turns page on calendar (8) to new day.

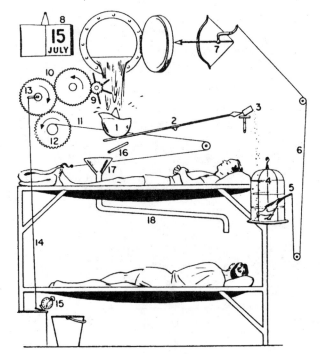

In the meantime, water falling over waterwheel (9) turns gears (10) which wind string (11) on drum (12). This pulls blanket up over Oscar. Arm (13) pulls on cord (14) and raises board under alarm clock (15) sliding same into bucket of water.

In case of mechanical breakdown at any point in the system, helmet is tipped by off-center peg (16) emptying water into funnel (17). Pipe (18) directs water onto Elmer, who is sleeping below. The theory is that Elmer will get up and do something about that open port—or about Oscar.

◆Objective

To apply your knowledge of simple machines to design and build a "Rube Goldberg" *tardiness preventer*. By helping you boil an egg for breakfast, this device will ensure that you always reach homeroom on time.

Materials

• an uncooked egg

• a pan of water

• any materials needed to build the device, provided they are not explosive, toxic, combustible, or in some other way dangerous to use

* From *Basic Machines,* prepared by the Bureau of Naval Personnel as Navy Training Course NAVPERS 10624-A. Reprinted under the title *Basic Machines and How They Work* by Dover Publications, New York (1971).

◆Procedure

1. Study the "Porthole Closer—Blanket Puller-Upper" Rube Goldberg device. It uses levers, pulleys, and an inclined plane to solve several different problems.

Your Tardiness Preventer does not have to be that elaborate, but you may wish to include some of the porthole closer's ideas in your design. The following requirements must be met:

• Your device must lower an uncooked egg a vertical distance of half a meter (minimum) into a pan of water without breaking the egg.

• You are not allowed to touch either the egg or the device once you set the device in motion.

• You must use a lever, a pulley, *and* an inclined plane as parts of the device.

2. Do not worry too much about getting the right answer. Wrong answers to this problem can be both educational and amusing. Testing designs that *may not* operate the way that you predicted is an excellent way to increase your understanding of how simple machines work.

3. Draw a rough sketch of your design in the space below.

GUIDE TO ACTIVITY 19

"Unsimple" Machines: Designing a "Rube Goldberg" Device

◆What is happening?

The best way to develop an intuitive understanding of how simple machines work is to experiment with actual machines. Every group that designs and tests a device, *whether or not* the device can lower the egg without breaking it, has successfully completed this activity.

The activity deliberately avoids asking for a numerical analysis of the forces and mechanical advantages of the components of the completed device. This activity is intended to be a practical, rather than numerical, investigation of simple machines.

The process of designing and building the device will improve students' understanding of simple machines. It also provides an opportunity to creatively apply the knowledge gained in previous activities.

The only wrong way to do this activity is to use some type of dangerous design or dangerous equipment. Make sure students are safety conscious; have them consider safety hazards that could result if their machines were to break or malfunction.

◆Time management

One class period (40–60 minutes) should be enough time to complete the activity and discuss the results.

◆Preparation

This activity can be used as a short project to be completed at home. It is a good idea to give students some time to work on the initial plan during class, however.

◆Suggestions for further study

What is the longest "Rube Goldberg" machine that your class can make? Divide into four- or five-member teams. Each team must create its own machine which includes a lever, a pulley, or an inclined plane. Each team must try to maximize the horizontal distance over which their machine acts. Since all the machines will be hooked together to make a super-long machine, one or two consultants from each team must talk with other consultants to decide how the machines will be "hooked" together.

◆Answer

The following drawing of a Tardiness Preventer is based upon a design created by Kate Ziegler and Carolyn Scarborough, teachers at Zebulon Middle School, Zebulon, North Carolina. Thanks, Kate and Carolyn, for a simple design *that works!*

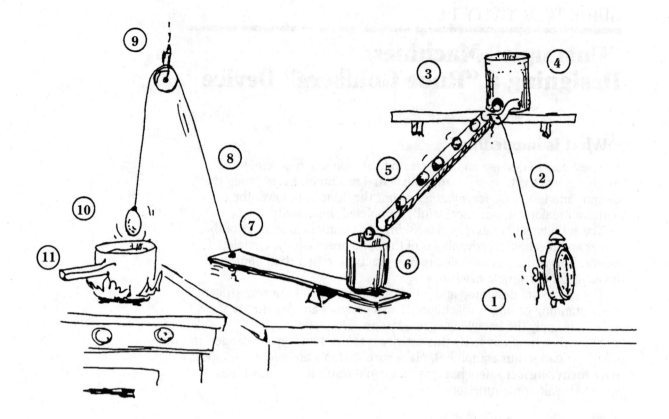

When the alarm clock (1) rings, the clock's key turns, winding up a string (2) that is tied to the key. The string pulls off a piece of tape (3) covering a hole in the bottom of a can of marbles (4). The marbles roll down the trough (5) into a lower can (6) that is attached to a seesaw (7). When the lower can is full of marbles, it pushes the arm of the seesaw down, allowing a string (8) that runs through a pulley (9) to lower the egg (10) into a pan of water (11) on the stove.

MODULE 5

Seesaws, Steering Wheels, and Screwdrivers: Applications of Torque

◆Introduction

• Why do buses and trucks have large steering wheels?

• Homerun hitters hold their baseball bats at the very end of the handle, but single hitters grip their bats closer to the center. Why?

• Why does using pliers make it easier to twist the top off of a bottle?

Every time you open a jar of jelly, steer a car, or close a door, you are applying something called **torque** to an object. Jar tops, steering wheels, and door handles all rotate about an axis when you apply a force to them. The product of this force and the perpendicular distance from where you apply the force to the axis of rotation is the torque. We write this

Torque = Force x distance = F x d.

Turning any object in a circle requires torque. From the equation for torque you can see that the amount of torque being applied to an object depends on two factors: the amount of force being applied and the distance from the fulcrum or axis of rotation to the point where the force is applied.

The activities in this module are designed to show different ways that we experience torque, and to help students refine their intuitive knowledge of torque.

◆Instructional Objectives

After completing the activities and readings for Module 5, students should be able to

• define torque [Activity 20]

• demonstrate examples of torques that we encounter in everyday life [Activities 20–24]

• calculate or estimate the amount of torque acting about an axis of rotation [Activities 20 and 24]

◆Activities

This module includes the following activities:

Activity 20: Mini Experiments on Torque

Activity 21: No Magic in the Witch's Broom

Activity 22: Walking Yo-Yo

Activity 23: Easing Up on Screwdrivers

Activity 24: Hammering Away at Torque

ACTIVITY 20 WORKSHEET

Mini Experiments on Torque

◆Background

Why is it easier to open a jar of jelly with a pair of pliers than with your bare hands? You can exert the same amount of force either way, but somehow the pliers make turning the lid easier. Any time you try to turn or rotate an object you are applying something called **torque** to the object. Torque depends on two factors: the force exerted and the perpendicular distance from where the force is applied to the point or line about which the object rotates. Torque in fact is the product of these two factors: Torque = Force x distance = F x d. So you can see that even if the force remains constant, increasing the distance from where you apply the force to the axis or point of rotation increases the torque. Now do you see how using pliers increases the torque that causes the lid to rotate? In this activity you will perform some mini experiments that illustrate the concept of torque.

◆Objective

To refine your intuitive knowledge of torque by investigating some common ways we experience torque

◆Procedure

Part I Heavy water

1. Place one cup in each hand. Hold one cup under your chin (as though you were about to drink). Hold the other cup at shoulder height, an arm's length away from your body (as though you were passing it up to someone). Keep the two cups in those positions for 60 seconds.

2. Which cup of water exerts more torque on your arm? Explain why.

_____

_____

Part II Hang ten

3. Good surfers can "hang ten" (walk to the nose of the board and hang their toes over). Why doesn't the board flip up and cause a wipe out when they do this trick? Your teacher will demonstrate how surfers accomplish this with the help of two volunteers.

4. Perform this demonstration with the board *placed close to the ground*. The surfer volunteer must weigh less than 70 kg (about 150 lbs). If the surfer who hangs ten loses his or her balance, the spotter volunteer should help steady him or her and prevent a fall. The teacher should be ready to catch the board in case it flips up.

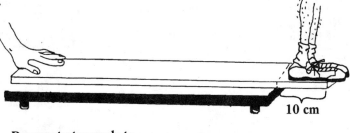

Demonstrator ready to catch end of board

10 cm

Spotter stands here

Materials

Each group will need

• two cups of water, both filled to the top

• a 2" x 6" plank 3 m long (this is the preferred size; the *minimum* size acceptable is a 2" x 4" that is 2.5 m long)

• access to a door

• two common bricks (with holes in them)

• a large dowel (it must fit through the holes in the bricks; the length should be approximately 0.75 m to 1.25 m)

• duct tape

Vocabulary

• **Torque:** A quantity that is a measure of the turning effect of an applied force; it is the product of the applied force and the perpendicular distance from the point of application to the axis or point of rotation of the object being turned. A torque is applied any time an object is turned or rotated.

! **Do not perform this activity if you do not have a board that meets the size requirement!**

! **The low support on which the board rests must be only a few centimeters above the ground!**

! **The surfer must weigh 150 lbs or less!**

! **Make sure someone is in a position to catch the board if it starts to slip!**

! **Make sure there is a spotter to steady the surfer!**

Place the board on a low support so that one end extends *10 cm* over the edge. Position the spotter beside the overhanging end of the board. Have the surfer move carefully onto the overhanging end of the board until his feet are both on the unsupported part of the board.

5. How is it possible for the board, which weighs less than the surfer, to support the surfer's weight?

_____

_____

_____

_____

Part III   Please close the door when you leave the room!

6. Try to close a door by placing *one finger close to the hinged edge of the door*. Next try closing the door by placing *one finger in the middle of the door*. Finally, close it by pushing *at the outer edge of the door.*

7. Which placement of your finger produces the greatest amount of torque for closing the door? Why?

_____

_____

_____

_____

! **Be careful not to hit anyone with the dowel or bricks while twisting the dowel. Make sure there is enough duct tape to prevent the bricks from falling off.**

! **Do not twist the dowel and bricks so rapidly that you lose control of them!**

Part IV   Twist and shout

8. Place the dowel through a hole in each brick. Wrap the duct tape securely around both ends of the dowel to prevent the bricks from sliding off.

Grasp the dowel in the middle, with a brick on either side of your hand. Move the bricks as close to your hand as possible. Twist the dowel and bricks.

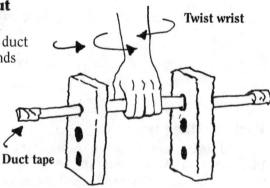

Twist wrist

Duct tape

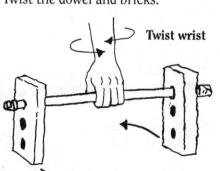

Twist wrist

9. Move both bricks out to the opposite ends of the dowel. Hold the dowel in the middle, and twist the bricks as before.

10. Describe how the torque required to twist the dowel and bricks changes when the position of the bricks is changed.

_____

_____

_____

_____

Part V   A seesaw solution

11. Children make use of an intuitive knowledge of torque every time they play on a seesaw, even if they can't explain it.

Suppose two children shift their positions along the arm of a seesaw until they discover that the light person sitting near the end of the seesaw can support the heavy person sitting near the middle.

12. If one child weighs 200 N, the other weighs 400 N, and the lighter child is sitting 2 m away from the center of the seesaw, can you use the equation _Torque = Force x distance_ to find out how far from the center the 400-N child has to sit to keep the seesaw level? (Hint: The center of the seesaw is the point of rotation for the torque exerted by each child. The seesaw will be level when the two torques are equal.)

GUIDE TO ACTIVITY 20

Mini Experiments on Torque

◆What is happening?

The purpose of this activity is to refine the intuitive knowledge that students already have of torque by having them think about some common experiences of torque in terms of Force x distance.

In Part I, a tongue-in-cheek explanation for why one arm gets tired before the other could be that one cup contains "heavy water" and the other contains regular water. (Heavy water has deuterium, a heavy isotope of hydrogen, in place of ordinary hydrogen.) Actually, the outcome can be explained in terms of torque.

The cup that is held away from the body is far from the point of rotation (the shoulder). Supporting the weight (F) at that distance (d) produces a great amount of torque (F x d). To balance the cup, the shoulder muscles have to exert a large counter torque; consequently, they tire quickly. With the cup held close to the chin, the short distance between the force and the point of rotation produces a small torque. Little counter torque is required, so the muscles do not tire as easily.

In Part II, although the board weighs less than the surfer, the length of the board from the edge of the table (its axis of rotation) to the opposite end is great. A little force (the weight of the board) times a large distance can produce the same torque as the surfer's weight times a short distance (the length of board hanging over the edge of the table—10 cm).

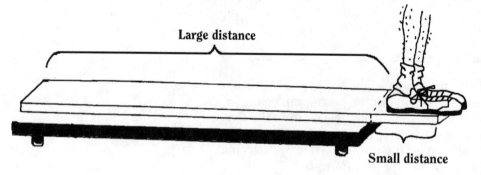

Large distance

Small distance

In Part III, the door closes faster when the force is exerted farther from the *axis of rotation* (the hinged edge). Assuming that the same force is exerted at each location, the forces acting at greater distances from the axis of rotation produce greater torques because torque is the product of force and distance. Although the torques are not measured directly, observing how quickly the door moves allows an indirect comparison of the magnitude of the three torques.

In Part IV, the combined mass of the dowel and bricks remains constant, but the torque required to twist them increases when the bricks are moved away from the point of rotation (the place where you hold the dowel). The greater the distance the bricks are from the axis of rotation, the more torque is required to twist them. (This is an example of *rotational inertia*, or the *moment of inertia*. Bodies that *are not rotating* [such as the bricks] tend to remain nonrotating; bodies that are rotating [such as a spinning top] tend to continue rotating. Rotational inertia depends on two factors: the mass of the body and the distribution of the mass relative to the axis of rotation.)

A discussion of the solution to Part V is included in the answers given below. You may wish to guide a class discussion of the solution after students have had an opportunity to try the solution themselves.

◆Time management

One class period (40–60 minutes) should be enough time to complete the activity and discuss the results.

◆Preparation

The five parts of this activity can be performed during a single class period. Alternatively, you may wish to perform the Mini Experiments as demonstrations and use them to stimulate a class discussion of torque. Another mode of presentation is to have students do one part of the activity per day for several successive days.

For Part II, set up the board before class or while the students are filling their cups with water for the Heavy Water activity. You can add to the drama of this demonstration by playing a record of some beach music. A 3 m long 2" x 6" is the *preferred size* of lumber for this demonstration. *Do not attempt* to do this demonstration unless you can obtain a 2" x 4" plank at least 2.5 m long. This is the *minimum* acceptable size. Smaller boards may flip up unexpectedly. Supervise this demonstration closely and be sure that your spotter remains attentive to the surfer at all times.

For Part III, if you have a lot of cabinets in your classroom, have all the students do this as a "fingers-on" experiment. Otherwise, demonstrate it, or perhaps assign it as a homework problem. Actually experiencing the relative ease or difficulty of closing the door while pushing from different locations is a good way to provide personal experience of torque.

Ideally, Part IV should be a hands-on experiment done in small groups. Each person should have the opportunity to twist the rod. Wooden dowels, broomsticks, plastic plumbing pipe, or metal ring stand rods can be used to support the bricks. Whatever type of support for the bricks you decide to use, be sure that duct tape or something else is placed on the ends of the rod so that *the bricks cannot slip off.* Caution students to stand back from the person twisting the bricks. The title of this activity is borrowed from a song performed by the Beatles. You may wish to introduce this activity by playing the song.

Part V is a pencil and paper activity that students can do as a homework assignment or after finishing the first four parts of the activity. If you happen to have a seesaw on the school grounds, take the class out and demonstrate how lighter people can hold up heavier people by adjusting where they sit.

◆Suggestions for further study

How would you use a plank and brick to lift your teacher a distance of three finger widths off the floor with the force you could exert with just three fingers? Where would the teacher stand? Where would the brick be placed? Where would you exert the three-finger force?

At a playground there is a seesaw that is constructed so that people can only sit at the ends of the seesaw. This seesaw, however, can be placed on the cross support at different locations. If you weigh more than the person you are balancing, would the cross support have to be closer to you or to the other person?

Assume that you had the job of getting a merry-go-round full of kids going and that you wanted to do the job with minimum effort. Would you have all the kids sit as far away from the middle as possible or would you have them crowd as close to the middle as possible?

◆Answers

2. Both cups exert about the same amount of downward force, because they weigh about the same. However, the cup held at arm's length exerts more torque because it is applying that force a greater distance from the point of rotation.

5. The surfer produces a torque by exerting a large force (his or her weight) at a point that is a short distance from the axis of rotation (the edge of the support). The board produces a larger torque by exerting a smaller force (its weight) a longer distance from the axis of rotation. We can tell that the board's torque is greater than the surfer's torque because the board is able to support the surfer. The board does not move on its axis of rotation.

7. The finger placed at the outer edge of the door produces the greatest torque because it is applying a force over the longest distance from the axis of rotation. The hinged edge is the axis of rotation.

10. The closer together (nearer the hand) the bricks are placed, the less torque is required. The torque decreases as the bricks move toward the center because the center of the rod is the axis of rotation. The shorter the distance from the axis of rotation, the less torque required to move the bricks.

12. The two torques are equal when

$$\text{Torque}_{\text{[large child]}} = \text{Torque}_{\text{[small child]}}, \text{ or}$$

$$F_{\text{[large child]}} \times d_{\text{[large child]}} = F_{\text{[small child]}} \times d_{\text{[small child]}}.$$

So, $400 \text{ N} \times d_{\text{[large child]}} = 200 \text{ N} \times 2 \text{ m}$.

Since the d in the equation is the distance from the point where the force is applied to the point of rotation, and the point of rotation is the center of the seesaw, the d is exactly what we want to know—how far away the large child is sitting from the center (the distance from the point of rotation to the point where the force is applied). So we can solve for d:

$$400 \text{ N} \times d_{\text{[large child]}} = 400 \text{ N} \cdot \text{m}$$

$$d_{\text{[large child]}} = 400 \text{ N} \cdot \text{m}/400 \text{ N}$$

$$d_{\text{[large child]}} = 1 \text{ m}.$$

So, the large child has to sit 1 m from the center in order to balance the seesaw.

ACTIVITY 21 WORKSHEET

No Magic in the Witch's Broom

◆Background

We have to remember that torque is the result of *two* factors—force *and* distance from the point of rotation—or we may be misled into some strange conclusions. Using screwdrivers, pliers, and wrenches does not make us stronger, and as we'll see in this activity, there is no magic in a witch's broom.

◆Objective

To investigate how the concept of torque explains an otherwise surprising result—a small force overcoming a larger force

◆Procedure

Materials

Each group will need
• a broom
• a paper cup or other small, unbreakable object

Part I   One-finger power house

1. Place the cup on the floor. Have one person turn the broom upside down and grab the handle near the fiber end with both hands. Have the other person kneel down and place *one finger* under the end of the handle. Place the end of the handle to one side of the cup.

The object for the person using both hands is to move the end of stick sideways, knocking over the cup. The person using one finger must try to prevent the cup from being tipped over.

The person using two hands *must*

• keep his hands close together below the fiber end

• *not* jerk or lift the broom off the other person's finger

• *not* move the handle up or down

Point of rotation

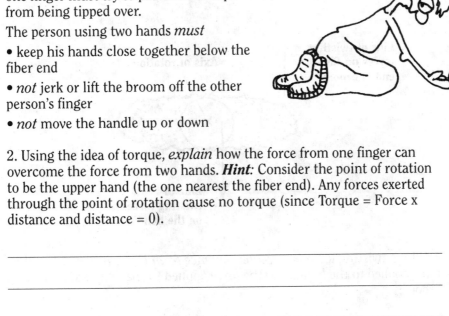

2. Using the idea of torque, *explain* how the force from one finger can overcome the force from two hands. **Hint:** Consider the point of rotation to be the upper hand (the one nearest the fiber end). Any forces exerted through the point of rotation cause no torque (since Torque = Force x distance and distance = 0).

_____

_____

_____

_____

Part II A slight twist

3. Have one person use both hands to grab the broom near the end of the handle. Have the other person use both hands to grab the broad fiber end of the broom. The two people should twist the broom in *opposite directions*. The axis of rotation runs lengthwise down the middle of the handle.

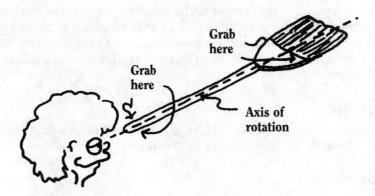

4. Who has the greater twisting strength: the person holding the handle, or the person holding the fiber end?

_____

5. You can use the concept of torque to explain why twisting the broom is easier for one of the experimenters. Answer questions 6 through 10 after studying the illustration below. The illustration shows the view that the person sighting down the broom handle (in the previous illustration) would see.

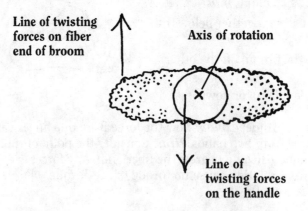

6. Which twisting force is a *greater distance from the axis of rotation*: the force applied to the *handle,* or the force applied to the *fiber end* of the broom?

_____

7. Assume that the force applied to the handle is the *same* as the force applied to the fiber end of the broom. Torque is equal to force times distance from the axis of rotation. Which force would produce the most torque: the force applied to the *handle*, or the force applied to the *fiber end* of the broom? Why?

_____

_____

8. The greater the torque being applied, the easier it will be to turn the broom. Which person will find it easier to turn the broom: the person applying the torque to the *handle*, or the person applying the torque to the *fiber end* of the broom? Why?

_____

_____

9. Suppose there are two brooms. The handles are identical, but one has a *broad* fiber end and the other has a *narrow* fiber end. If you were asked to select a broom and grab the fiber end for a twisting contest, which broom would you choose: the one with the narrow fiber end, or the one with the broad fiber end? Why?

_____

_____

10. Now apply the same ideas and use the concept of torque to explain why buses and trucks have large steering wheels.

_____

_____

_____

_____

GUIDE TO ACTIVITY 21

No Magic in the Witch's Broom

◆What is happening?

Torque is the product of force and the distance from the axis of rotation. In the first part of this activity, "One-finger power house," a small amount of force is applied by one finger at the end of the broom. The one finger is able to produce a *large* amount of torque because the force is applied a great distance from the point of rotation.

In most cases, the person holding the fiber end of the broom with both hands cannot overcome the torque produced by the single finger. A hand can exert much more force than a single finger, but when the hands are located almost on top of the axis of rotation, the torque they produce is quite small—smaller than the torque of the single finger.

For Part II, "A slight twist," both people will be able to exert about the same amount of force. However, since the person holding the fiber end of the broom is exerting that force a greater distance from the axis of rotation, the torque produced on the fiber end of the broom is larger than the torque produced on the handle end. So, the person holding the fiber end can overpower the other person easily.

◆Time management

One class period (40–60 minutes) should be enough time to complete the activity and discuss the results.

◆Preparation

You may wish to have students provide the brooms that are used in this activity.

◆Suggestions for further study

Try the "One-finger power house" activity again and have the one finger move up the handle to the point where the two hands can just overcome the finger. At that point, the torque created by the hands and the counter torque created by the finger are almost equal. The number of hand-widths from the top hand to the finger position is about how many times stronger a hand is than a finger.

Assume someone challenges you to a twisting contest similar to the one described in "A slight twist." You accept the challenge providing you can choose the broom and the end at which you do the twisting. Would you choose a narrow kitchen broom or a very wide custodial push broom for the contest? Use ideas of torque to explain your selection.

◆Answers

2. The person holding the fiber end of the broom produces little torque. The broom's point of rotation is the top hand of the person holding the fiber end. That person's lower hand exerts its force only a short distance away from the point of rotation. Therefore, even though the force exerted by the lower hand is large, the distance is small and therefore only a small torque is produced.

The finger at the end of the broom, however, is a large distance from the point of rotation. Consequently, even though applying a small force, the finger produces a large amount of torque.

4. The person holding the fiber end of the broom will always be able to overpower the person holding the handle.

6. The force applied to the fiber end of the broom is a greater distance from the axis of rotation (the axis runs down the center of the handle).

7. A force applied to the fiber end will produce more torque than an identical force applied to the handle, because the force on the fiber end acts through a greater distance.

8. Since the person grabbing the fiber end is producing more torque (question 7), he or she will find it easier to turn the broom.

9. The broom with the broad fiber end should be chosen because the force can then be applied a greater distance from the axis of rotation, producing a greater torque and consequently making it easier to turn the broom.

10. Large buses and trucks are heavy and have large wheels that are difficult to turn. Turning those wheels requires the application of a large amount of torque to the steering column by the steering wheel. Torque can be increased by applying a larger force, or by applying the same force over a larger distance. A large steering wheel increases the distance through which the driver's force acts. This allows the driver to apply a moderate amount of force to the steering wheel and to produce a large amount of torque to turn the front wheels of the truck.

ACTIVITY 22 WORKSHEET

Walking Yo-Yo

◆Background

Walking a yo-yo produces some surprising results—unless you understand the concept of torque. Can you apply the concept of torque to predict which way a yo-yo will roll when "walked?"

Each group will need
• a yo-yo

◆Objective

To apply the concept of torque to understand the behavior of a walking yo-yo

◆Procedure

1. Place the yo-yo on a flat (nonslippery) surface so that the string comes off the *top* of the axle. Can you predict how it will move if you gently pull the string parallel to the top of the table?

The yo-yo will move (check one):

_____ in the same direction as the pull.

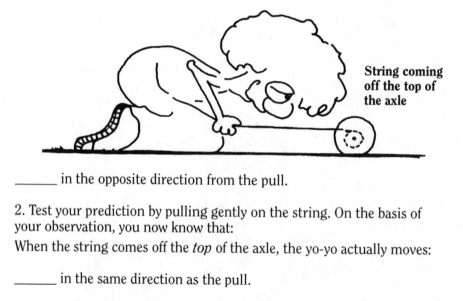

String coming off the top of the axle

_____ in the opposite direction from the pull.

2. Test your prediction by pulling gently on the string. On the basis of your observation, you now know that:

When the string comes off the *top* of the axle, the yo-yo actually moves:

_____ in the same direction as the pull.

_____ in the opposite direction from the pull.

3. Now turn the yo-yo over so that the string comes off the *bottom* of the axle.

Can you predict how it will move if you gently pull the string parallel to the top of the table?

The yo-yo will move:

_____ in the same direction as the pull.

_____ in the opposite direction from the pull.

String coming off the bottom of the axle

4. Test your prediction by pulling gently on the string. On the basis of your observation, you now know that:

When the string comes off the *bottom* of the axle, the yo-yo moves:

_____ in the same direction as the pull.

_____ in the opposite direction from the pull.

5. Can you apply the concept of torque to explain your results?

_____

_____

_____

_____

_____

_____

GUIDE TO ACTIVITY 22

Walking Yo-Yo

◆What is happening?

When the yo-yo's string is pulled, two forces are affecting its motion: the force exerted by the string and the opposing force of friction. When the yo-yo is moving at a constant rate, these forces are equal in magnitude and opposite in direction. (If the forces were *unequal* there would be a net force acting on the yo-yo. By Newton's second law, $F = ma$, this means that the yo-yo would be accelerating—it would not be moving at a constant rate.)

Even though the opposing forces exerted by the string and friction are equal, *the torques that these forces produce are not equal*. Remember, torque is equal to force times *distance from the axis of rotation*. The yo-yo's axis of rotation runs through the center of the yo-yo.

Let's see how the idea of torque can be applied to explain why the yo-yo behaves as it does. First, consider whether the forces are causing *clockwise* or *counterclockwise* torques. Torques in the same direction add; those in the opposite direction subtract or oppose each other.

When the string comes off the top of the axle, the torque produced by the string and the torque produced by the friction are both counterclockwise. (See diagram below.)

Since the torques are both counterclockwise, the effects of the torques add together to rotate the yo-yo in the direction of the pull (to the left in the diagram).

Contrary to what most people expect, the yo-yo also rolls in the direction of the pull when the string comes off the bottom of the axle. Why?

First, as shown in the diagram below, the force from the string produces a *clockwise* torque. The frictional force produces an opposing *counterclockwise* torque.

The yo-yo will rotate in the direction of the greater torque. The frictional force is acting at a *greater distance* from the point of rotation than is the pulling force. Since these two forces are equal, the frictional force will produce the greater torque because it is acting over a longer distance. The yo-yo will then rotate in the *counterclockwise direction* (the same direction as the torque produced by the frictional force) and move in the direction of the pull.

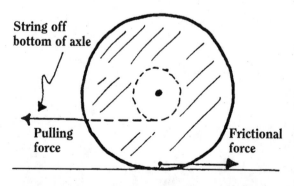

Regardless of whether the string comes off the top or bottom of the axle, the yo-yo moves in the direction of the pull.

◆Time management

Half of one class period (20–30 minutes) should be enough time to complete the activity and discuss the results.

◆Preparation

Before performing this activity, check to be sure that the yo-yos will balance on their edges and roll smoothly. Certain types of yo-yos tip over when you pull the string. An alternative is to make your own inexpensive yo-yos using Tinker Toys™ and string.

◆Suggestions for further study

Assume you are in the business of making yo-yos and have a research lab for testing out new ideas. Your researchers have made two identical yo-yos with the exception that one has a fat axle around which the string is wound and the other has a skinny axle. If the yo-yos are wound up and the strings are pulled with the same force, which yo-yo should begin spinning faster?

◆Answers

2. The yo-yo moves in the same direction as the pull when the string comes off the top of the axle.

4. The yo-yo moves in the same direction as the pull when the string comes off the bottom of the axle.

5. See the "What is happening?" discussion.

ACTIVITY 23 WORKSHEET

Easing Up on Screwdrivers

Materials

Each group will need

• several average-size screws

• a board with 6 or more holes drilled into it (the screws should fit the holes tightly)

• 2 identical, average-size screwdrivers

• masking tape

• 2 small blocks of wood (about 2 cm x 2 cm x 5 cm)

◆Background

Rotating any object requires the application of torque to the object. Increasing either the *force* applied to the object or the *distance* from the point of application to the point or axis of rotation also increases the torque. Can you use this information to build a better screwdriver?

◆Objective

To apply the concept of torque in order to design a screwdriver that can get a screw into a piece of wood more easily; i.e., with less force

◆Procedure

1. There is only one restriction on the use of the available materials in your design: You can only use *one* of the screwdrivers in your invention; the other screwdriver is to be used to make comparisons.

2. Make a drawing of your invention.

3. Build your improved screwdriver and compare its performance to that of the ordinary screwdriver. Use the concept of torque to explain why your new improved invention makes it easier (requires less force) to get the screw into the wood. Recall that torque applies to lines or axes of rotation as well as points of rotation.

_____

_____

_____

_____

_____

_____

4. Why do you think that manufacturers haven't used your invention? After all, your invention does reduce the amount of effort required to turn the screw into the wood.

_____

_____

_____

_____

_____

_____

GUIDE TO ACTIVITY 23

Easing Up on Screwdrivers

◆What is happening?

The most common solution to the challenge is to tape the block or blocks of wood to the handle of the screwdriver. This increases the diameter of the screwdriver considerably. Why does increasing the diameter of the handle make the effort less?

Assume all the screws require the same amount of torque to turn, regardless of the screwdriver or its handle. A small force coupled with a long distance can produce the same torque as a larger force coupled with a smaller distance:

$$F_{[large]} \times d_{[small]} = F_{[small]} \times d_{[large]}$$

When you turn a screwdriver, you wrap your hand around the handle and exert twisting forces all around the handle. To simplify matters, let's combine those forces into one twisting force exerted on the handle. Since we are turning the screwdriver about an axis, we must look at the distance from where this force acts to the axis. The following diagram shows where the twisting force is exerted on a screwdriver with a small handle.

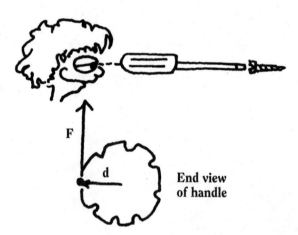

End view
of handle

This diagram shows where the twisting force is applied in relation to the axis of rotation on a screwdriver with a larger handle.

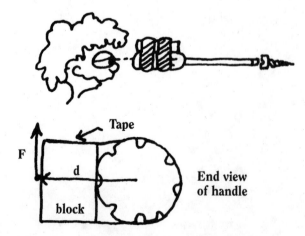

End view
of handle

So, making a larger handle increases the distance term in the equation for torque (Torque = Force x distance), allowing you to exert a smaller force and still produce the same torque.

◆Time management

Half of one class period (20–30 minutes) should be enough time to complete the activity and discuss the results.

◆Preparation

If you ask your students to bring in the screwdrivers required for this activity, be sure to specify straight blade (*not a* phillips blade) screwdrivers. Otherwise, you will need to have two types of screws available.

If you do not have a drill to make starter holes for the screws, you can make holes by hammering a nail part way into a board, then pulling it out. Just be sure that the nail is a size that will make a hole that the screw can fit into snugly. Have the pieces of wood with starter holes ready before class, for both time and safety reasons.

◆Suggestions for further study

In the activity "Easing Up on Screwdrivers," you changed the screwdriver so you could exert minimal force to produce the torque necessary to get the screw into the wood. How might you redesign the head of the screw so that the screw could be more easily driven into the wood? You might also have to make modifications in the blade of the screwdriver.

If you sold replaceable blades for screwdrivers and wanted to increase your sales, would you suggest that the screwdriver handles be made fatter or skinnier?

◆Answers

2. Most people meet the challenge by taping a block or blocks of wood to the handle, increasing its diameter.

3. With a larger handle, the distance from the point of the force's application to the axis of rotation is greater, allowing a smaller twisting force to produce the same torque. A smaller required force means that removing the screw is easier.

4. Some heavy screwdrivers *do* have large handles. However, there are some practical limitations on the sizes of screwdriver handles. The size of the handle is matched to the amount of torque that the blade can withstand. A small screwdriver blade might not be able to take the greater torque produced by a larger handle without breaking. If manufacturers put large handles on screwdrivers with small blades, we would be breaking a lot of screwdrivers.

Another limitation is the space in which screwdrivers sometimes have to fit. In a corner or cramped space, there may not be room to fit or turn a large-handled screwdriver.

Hammering Away at Torque

Materials

Each group will need
- a hammer
- a 35-cm length of 2" x 4" lumber
- 2 nails (6d or larger common nails)
- 2 heavy-duty rubber bands
- a medium or medium-small rubber band
- a metric ruler or meter stick
- **optional:** a 5-cm length of masking tape

◆Background

Why is it easier to pull a nail out of a piece of wood with a hammer than with your hands? Applying the concept of torque can not only answer this question, but also shows how best to use a hammer—where to push on its handle in order to use the least amount of effort to pull out the nail.

◆Objective

To use the concept of torque in order to determine the best way to use a hammer

◆Procedure

1. You will simulate nail pulling by pulling against rubber bands attached to the board.

Make the attachment points for the heavy-duty rubber bands by hammering the two nails part way into the corners of one end of the board.

Loop the heavy-duty rubber bands over the nail heads and over the claws of the hammer. These will be the *resistance* rubber bands.

You will use the stretch length (the distance from the handle to the end of the band) of the small rubber band (the *effort* band) to measure the force needed to pull a nail. The nail is pulled out when the hammer lifts off the wood at point X.

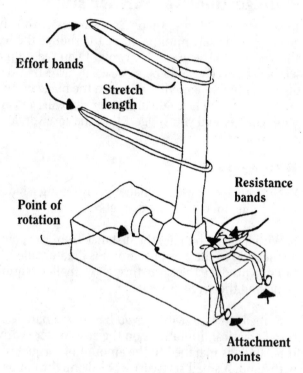

2. Think about the following questions while doing the experimental procedures that follow:

• Where should you push on the handle of a hammer to exert the least effort in pulling out a nail?

• Where should the nail be placed for the least effort—close to the handle (at point C) or out at the end of the claw (at point E)?

• Can you use the concept of torque to explain your observations?

3. Place the resistance bands at point C on the claw (near the handle).

Draw a line on the board to mark the point of rotation, D.

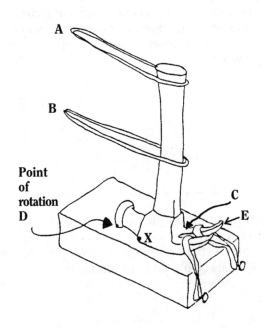

A

B

Point
of
rotation
D

C

E

X

Place the effort band at point A (the end of the handle). Pull the effort band *parallel to the table top*.

Measure the *stretch length* of the effort band just when point X lifts off the wood. The stretch length is the distance from the handle to the end of the band.

Slide the effort band down to point B.

Pull the effort band parallel to the table top. Measure the stretch length of the effort band as you did before, just when point X lifts off the wood.

Stretch length for pull at point A = _____ cm.

Stretch length for pull at point B = _____ cm.

4. Place the resistance bands at the end of the claw (point E). (You may have to use masking tape to hold the bands at the ends of the claw.)

Place the effort band back at point A (the end of the handle). Align the striking surface of the hammer with the reference line D on the board. Pull the effort band *parallel to the table top*.

Measure the stretch length of the effort band just when point X lifts off the wood.

When the resistance bands are at point E:

Stretch length for pull at point A = _____ cm.

Stretch length for pull at point B = _____ cm.

5. When the resistance bands were moved out to the end of the claw, how did the effort force (as indicated by stretch length) change?

_____ Pulling a nail required *more* effort. The stretch length *increased*.

_____ Pulling a nail required *less* effort. The stretch length *decreased*.

6. Which placement of the resistance bands and the effort band required the *least* amount of effort (as indicated by stretch length) to pull out a nail?

_____ Placing the resistance bands at C and the effort band at A.

_____ Placing the resistance bands at C and the effort band at B.

_____ Placing the resistance bands at E and the effort band at A.

_____ Placing the resistance bands at E and the effort band at B.

7. In order to *minimize the effort* needed to pull out a real nail from a board, you must place your hand on the hammer so that it can exert a *maximum amount of torque* on the nail. Where would you place your hand to maximize the torque? Why?

_____

_____

_____

_____

8. A nail stuck in a board exerts a torque on the hammer. In order to *minimize the effort* needed to pull out the nail, the claw of the hammer should be placed so that the nail exerts a *minimum amount of torque* on the hammer. Where should the claw be placed to minimize the nail's torque? Why?

_____

_____

_____

_____

GUIDE TO ACTIVITY 24

Hammering Away at Torque

◆What is happening?

Students probably discovered that pulling the nail requires the least amount of effort when the nail is closest to the handle and when you pull on the *end* of the hammer's handle. (In other words, the effort is least when the resistance bands are at point C and the effort band is at point A.)

The "nail" exerts the same amount of "wood-holding force" regardless of where it is placed on the claw, and regardless of where the hammer's handle is pulled. However, the *torque* that the nail exerts depends on the distance of this wood-holding force from the axis of rotation. The torque that the effort force exerts on the *hammer* depends on where that force is exerted on the handle.

You can minimize the effort required to pull out the nail by holding the handle so that *you* apply a *maximum amount of torque to pull it out*, and by placing the claw so that the *nail* exerts a *minimum amount of torque resisting being pulled out*.

Changing the position of the claw, or changing the point at which force is applied to the handle, changes only the amount of torque produced. It *does not change the force* exerted by the hammer or the nail.

Using rubber bands makes interpreting your nail pulling observations easier, because the resistance rubber bands always exert the same force against the hammer. When you are pulling real nails out of a board, each nail requires a different amount of force to remove because of irregularities in the board and the nails.

While force is not measured directly in this activity, the stretch of the effort rubber band is easily observed by students and offers a reliable method for comparing effort forces.

◆Time management

One class period (40–60 minutes) should be enough time to complete the activity and discuss the results.

◆Preparation

Test the rubber bands to be used in this activity. The torques produced must be large enough so that changes in rubber band length can be easily measured. You may want to drive the nails into the boards before class to save time and to avoid any possibility of injury to students.

◆Suggestions for further study

In the activity "Hammering Away at Torque," you were requested to keep the effort band parallel to the top of the table. What happens to the effort force when the effort band is not held parallel to the top of the table? Place the effort band at position A and measure the stretch length (effort force) needed to lift point X off the board with the resistance bands at point C. Now angle the effort band upward (45°) and see how much effort force (stretch length) is needed to lift point X. You may have to tape the effort band in place on the handle.

Try angling the effort band downward (45°) and see what happens to the effort force (stretch length). Complete the sentence: When the effort force is perpendicular to the handle (parallel to the table), the required effort is_____(*more, the same,* or *less*) than an effort force which is angled up or down.

Explain why it might be easier to pull out nails with a long-handled crow bar than a short-handled hammer.

◆Answers

Note: The numerical values you obtain for this activity depend on the rubber bands and hammer that you use. The following are intended as a guide only. *They are not the only correct and acceptable answers for this activity.*

3. Stretch length for pull at point *A* = *6 cm.*
   Stretch length for pull at point *B* = *8 cm.*

4. Stretch length for pull at point *A* = *9 cm.*
   Stretch length for pull at point *B* = *14 cm.*

5. When the resistance bands were moved out to the end of the claw, pulling a nail required *more* effort. The stretch length *increased.*

6. Placing the resistance bands at *C* and the effort band at *A* requires the *least* amount of effort to pull out a nail.

7. Grasping the end of the hammer's handle minimizes the effort required to pull the nail. Grasping it near the end allows the maximum possible distance from the point of force application to the axis of rotation, producing the maximum amount of torque for pulling the nail (Torque = Force x distance).

8. The nail exerts a minimum amount of torque when it is placed at point *C*, near the handle. The distance from the point of application of the nail's resistance force to the axis of rotation is minimized when the nail is placed at that location; the smaller the distance, the less the torque.

MODULE 6

Center of Gravity

◆Introduction

• Why are football linemen in a four-point stance hard to knock over?

• Do all circus high-wire performers have extraordinary balancing skills, or are "death-defying" tricks possibly easier than they look?

• Why do you begin to fall when you lean too far forward?

Most people's knowledge of **center of gravity** is based on their practical experiences with balancing things. This module illustrates everyday applications of the concept of center of gravity and shows how changing an object's center of gravity can allow it to behave in "mysterious" ways.

The concept of center of gravity can be defined in several ways. Depending upon the type of behavior we are trying to explain, one definition may be more useful than another:

• An object's center of gravity is the point at which, if supported in any orientation at this point, the object balances.

• An object's center of gravity is the point at which the entire weight of the body may be considered to be concentrated.

• An object's center of gravity is the point through which the force of gravity for the entire object may be considered to act regardless of the object's orientation.

The first definition corresponds to most people's practical concept of center of gravity. The second and third definitions are extremely important to physicists. Newton's Law of Gravitation could not have been formulated without using these ideas. The three definitions, of course, are different ways of saying the same thing.

The activities in this module are designed to illustrate both the formal and intuitive uses of the concept of center of gravity.

◆Instructional Objectives

After completing the activities and readings for Module 6, students should be able to

• define center of gravity [Activity 25]

• demonstrate how the location of an object's center of gravity affects the way that the object sits, stands, or moves [Activities 26, 28, and 29]

• experimentally determine where an object's center of gravity is located [Activity 27]

• balance an object by combining the concepts of torque and center of gravity [Activity 30]

Activities

This module includes the following activities

Activity 25: Center of Gravity—Let's Get Intuit(ive)

Activity 26: Discovering Your Personal Center of Gravity

Activity 27: Cynthia's Hanging

Activity 28: Center of Gravity: Magical Motivational Demonstrations

Activity 29: Motorcycle Michael and Balancing Belinda's Circus Act

Activity 30: Clip and Trim

ACTIVITY 25 WORKSHEET

Center of Gravity—Let's Get Intuit(ive)

Vocabulary

• **Center of gravity:** The point through which the force of gravity for an entire object may be considered to act, regardless of the object's orientation. In other words, the center of gravity is the point where we may consider all of an object's weight to be located.

◆Background

Understanding football, car racing, the pains of pregnancy, and the Leaning Tower of Pisa requires understanding an important concept in mechanics: **center of gravity**. A practical way of thinking of the center of gravity of an object is to consider it the point from which you could suspend the object and it would balance, no matter what the object's orientation is. This activity will help you see how the concept of center of gravity is important when trying to explain some everyday events.

◆Objective

To use the concept of center of gravity to answer some questions about everyday events

◆Procedure

1. Football linemen crouch low while blocking and tackling. To catch a pass, a wide receiver may have to leap and perform a pirouette.

Why is the lineman in a low crouch harder to knock over than the receiver performing a pirouette? (Assume that both players weigh the same.)

_____

_____

_____

_____

_____

2. The cars shown below are very similar—they both weigh the same and have equally powerful engines. However, one of the cars is a high rider and the other is a low rider.

Which of these two cars would probably win a race on a curvy racing track? Why?

_____

_____

_____

_____

_____

3. Pregnant women often complain of lower back pain.
Why does back pain accompany pregnancy?

_____

_____

_____

_____

_____

4. The Leaning Tower of Pisa is one of the most famous buildings in the world.
Why doesn't the Leaning Tower of Pisa fall over?

_____

_____

_____

_____

_____

GUIDE TO ACTIVITY 25

Center of Gravity—Let's Get Intuit(ive)

◆What is happening?

Grabbing students' interest and making them *want* to learn about the topic of the day's lesson is a never-ending challenge to teachers. This activity helps answer the students' question "Why do we have to study center of gravity?" by showing practical applications of this concept in everyday life.

◆Time management

Half of one class period (20–30 minutes) should be enough time to complete the activity and discuss the results.

◆Preparation

A good way to present this information to your class is to make overhead projection transparencies of the illustrations. Show the pictures, and use them as a basis for an informal discussion with the class, rather than having them write answers to the questions. Encourage students to describe their own experiences related to center of gravity. You may elicit some funny stories about "the time I fell off the. . . ."

◆Suggestions for further study

Use real people to try out some of the demonstrations illustrated in this activity. Have a person stand on one foot and see how much effort is needed to tip them over. Have that same person get down on all fours into a football stance. Now see how much effort is needed to tip the person over so they fall backwards. How much effort is needed to tip the football player over sideways?

To simulate pregnancy, hang three five-pound bags of sugar or flour in the front of a person and see how far they can bend over forward without moving his or her feet or losing balance. How far can he or she lean over without the bags of sugar (after pregnancy)? Recall that an object must fall over when its center of gravity gets outside its base of support.

Assume you wanted to keep the Leaning Tower of Pisa leaning a little bit longer. You could do that by changing the center of gravity of the tower. To change the center of gravity of the tower in the desired way, you would add weight to which side of the tower—the side opposite the lean or the side to which the tower leans?

◆Answers

1. In order to cause an object to fall over, the object's position has to be changed so that its center of gravity is not over the object's base of support. As the previous figures showed, the player in the low crouch with widely spaced points of support has a low center of gravity and must be raised and moved through a long distance before his center of gravity is outside his base of support and he falls. Therefore, it is hard to knock the lineman off his feet.

The center of gravity of the player in the pirouette position is much higher than the lineman's. He needs to stretch to reach the ball. But once he catches it, an onrushing defender colliding with the pirouetting pass-catcher only has to move the pass-catcher's center of gravity a small distance before it is outside his base of support and he falls.

2. Modern racing cars are low-slung and have widely-spaced wheels. They are designed this way so that they will have low centers of gravity. If all other factors are equal, a car with a lower center of gravity will be more stable going around sharp corners because it would have to tip severely before its center of gravity would be outside its base of support. The high rider's center of gravity would be outside its base of support after only tipping slightly. The higher stability of the low rider allows it to go faster without flipping over. The low rider should therefore win the race.

3. As the fetus grows larger, the pregnant woman's center of gravity shifts forward. If the woman's center of gravity moves in front of her toes, she will fall over forward. To prevent this from occurring, the woman instinctively arches her body backward. By unconciously changing the way that she stands as her midsection expands, a pregnant woman is able to keep her center of gravity directly above her feet. However, constantly arching her back causes her back muscles to fatigue, resulting in back pain.

4. The Leaning Tower of Pisa is still standing because its center of gravity is above the foundation supporting it. However, the ground under the tower is sinking unevenly, and the tower is leaning more and more with each passing year. Unless something is done to stop the increasing leaning, the tower will fall when its center of gravity is no longer above its foundation.

ACTIVITY 26 WORKSHEET

Discovering Your Personal Center of Gravity

◆Background

Human beings are top-heavy. When we stand, our center of gravity is located well *above* our feet. Walking and standing are possible only when the feet are underneath the body's center of gravity. This activity demonstrates what happens when the feet are not underneath the body's center of gravity.

◆Objective

To explore how the center of gravity in our bodies affects the way we stand, sit, and move

◆Procedure

Part I   Up against the wall

1. Stand facing a wall. Place your toes two foot-lengths away from the wall. (Note: Use *your own* feet to measure this distance, not an English ruler calibrated in feet and inches.)

Lean forward so that your head touches the wall. Straighten your back, and clasp your hands behind your back.

Without bending at the hips or pushing off the wall with your head, straighten into a standing position.

Women can often perform this task when men cannot. Why?

_____

_____

_____

_____

Part II   Back against the wall

2. Stand with your back touching a wall. Your heels should also be in contact with the wall.

Quickly bend over and touch your toes *without bending your knees*.

When you bend to touch your toes, how does the location of your center of gravity change?

_____

_____

_____

_____

Materials

Each group will need
• access to a wall
• a chair with a back

2 foot-lengths

Part III · Nose against the wall

3. Stand facing a wall. Both your nose and your toes must touch the wall. Stand on tiptoe while keeping your nose in contact with the wall.

When you try to stand on your toes, how does the location of your center of gravity change?

_____

_____

_____

_____

Part IV   Will you please rise

4. This challenge requires two people: a sitter and a holder.

The sitter should sit comfortably in a chair by leaning back, legs outstretched and arms folded across the chest in a restful posture.

The holder should place his or her little finger on the forehead of the sitter.

Now have the sitter try to stand without using his or her arms and feet.

How does the location of your center of gravity change when you move from a sitting position to a standing position?

_____

_____

_____

_____

GUIDE TO ACTIVITY 26

Discovering Your Personal Center of Gravity

◆What is happening?

This activity offers another way of grabbing students' interest and making them *enjoy* learning about their own center of gravity. Students benefit from actual physical experience of how the location of a body's center of gravity changes as the shape of the body changes. They also gain first-hand knowledge of how stability is affected by the location of the center of gravity relative to the base of support.

◆Time management

Half of one class period (20–30 minutes) should be enough time to complete the activity and discuss the results.

◆Preparation

Having the whole class do the demonstrations simultaneously may get noisy, but it will also promote interest in and conversation about centers of gravity.

◆Suggestions for further study

Get an empty cereal box and tape heavy weights (washers or bolts) to the inside so that the center of gravity of the box is not at the center of the box. Challenge others to move the box around *without lifting it* to find its center of gravity. (Hint: See where the box begins to tip over when pushed out over the edge of a table.)

Hang five-pound bags of sugar or flour on a person so they can easily perform the demonstration Head Against the Wall.

◆Answers

1. Men tend to have higher centers of gravity than women because men's upper bodies are larger and heavier than women's. When a man's higher center of gravity moves forward of his toes the downward force of his weight creates a torque around the axis of rotation that passes through his toes. The torque causes his body to rotate around this axis; i.e., he leans forward and his head is kept pressed against the wall. When a woman leans against the wall her center of gravity (located close to the ground) remains above her toes and no torque is created (Torque = Force x distance = Weight x 0 = 0).

2. The challenge posed is impossible to perform. You will fall forward before you can touch your toes.

Ordinarily, as you lean over and touch your toes, your posterior moves backward beyond your heels. With such a motion, your center of gravity remains *above your feet* and you do not topple over (the torques acting on you are balanced).

However, when you start with your heels against the wall and then lean over, the wall prevents your posterior from moving backward. Without the backward motion, your center of gravity moves forward of your toes; the weight acting through the center of gravity creates a torque that topples you forward.

3. This challenge is impossible for most people to perform. With your nose touching the wall, you cannot shift your center of gravity far enough forward to allow you to stand on your toes.

When standing still, your center of gravity is centered above your ankles. In order to stand on tiptoe you must bend forward slightly to shift your center of gravity toward your toes.

4. The sitter should find it impossible to rise without moving his or her feet and arms. One finger placed on the forehead of the sitter will keep the sitter seated.

When the sitter leans back, his or her center of gravity is well behind the feet. The only way the sitter can stand is to get his or her center of gravity over the feet. The finger and the folded arms prevent this.

When rising from a chair, people usually move their feet under the chair so that their feet are under their center of gravity. Shifting one's center of gravity when changing position is automatic because we do it so often.

ACTIVITY 27 WORKSHEET

Cynthia's Hanging

◆Background

If we hang Cynthia—or at least her picture—can we predict which way she will face? Cynthia, like every object, has a center of gravity; we can consider the force of gravity for the entire object to be acting only through this one point. Can we use this information to predict how Cynthia will hang when suspended from different points?

◆Objective

To use the concept of center of gravity to predict how an object will be oriented when suspended from different points

◆Procedure

1. Cut out Cynthia along the dashed lines. Punch three small, neat holes in Cynthia at points *A*, *B*, and *C*. (Cynthia's picture is on following page.)

2. Straighten out one end of a paper clip. Tape the clip to the desk top so the straight part sticks out over the edge.

3. Tie the lead weight to one end of the thread. Tie a loop in the opposite end of the thread. Hang this loop over the straightened paper clip so that the weight can swing freely below the tabletop.

4. Hang Cynthia on the straight paper clip by sticking the paper clip through her at point *A*. Cynthia should be free to rotate on the paper clip.

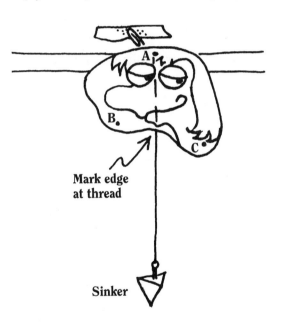

Mark edge at thread

Sinker

5. Mark where the thread comes out from behind Cynthia.
Remove Cynthia from the paper clip and draw a straight line from point *A* to the mark that you made.

6. Hang Cynthia from point *B* and repeat step 5.

7. Hang Cynthia from point *C* and repeat step 5 again.

Materials

Each group will need
• a lead weight (a 1-oz fishing sinker)
• 2 standard paper clips
• 50 cm of sewing thread or very light string
• scissors to cut out Cynthia
• a straight edge and pencil
• 10-cm length of masking tape
• a copy of the Cynthia pattern

8. There is a pattern to the way the lines are drawn. No matter where you hang Cynthia, something remains the same. What is it?

_____

_____

_____

_____

9. Slide the second paper clip over Cynthia's nose. Repeat steps 4–7.

10. What has changed about the way that the lines are drawn? No matter where you hang Cynthia with the paper clip on her nose, something remains the same. What is it?

_____

_____

_____

_____

11. Can you now predict how Cynthia will hang from any point on her surface?

_____

_____

_____

_____

_____

_____

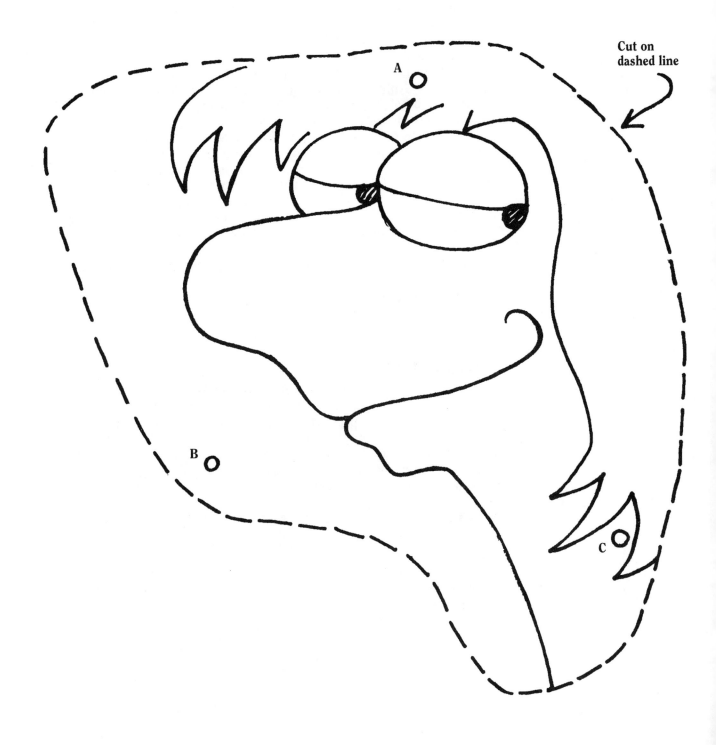

Cut on
dashed line

GUIDE TO ACTIVITY 27

Cynthia's Hanging

◆What is happening?

Every object has a point called the center of gravity, the point where the force of gravity *appears* to act.

If you carefully mark and draw the lines and if Cynthia and the weight are both free to rotate, the three lines will cross at a single point (or very nearly so). When you attach the paper clip to Cynthia's nose, the lines converge at a different point.

The point where the lines converge is Cynthia's center of gravity. No matter where Cynthia is suspended, the center of gravity is always directly below the point of support. As the lines show, adding the paper clip to Cynthia's nose changes her center of gravity.

In reality, the force of gravity acts simultaneously at all points of an object. However, physicists studying gravity's effects on the motion of an object may simplify the analysis by assuming that the center of gravity is the only point at which the force of gravity acts. Assuming that the force of gravity acts only at an object's center of gravity allows the physicist to consider just one gravitational force acting at one point rather than many little gravitational forces acting at all points in the object.

The center of gravity need not be in the object. For example, the center of gravity for a donut would likely be in the air at the center of the hole.

◆Time management

Half of one class period (20–30 minutes) should be enough time to complete the activity and discuss the results.

◆Preparation

Before class, make a photocopy of the Cynthia pattern on page 131 for each student group. Accurately determining a center of gravity may be more difficult when the paper clip is placed on Cynthia's nose. The paper clip exerts a torque about the point of support and causes Cynthia to rotate. However, since the paper clip has a long, thin shape, it exerts slightly differing amounts of torque depending on whether it is oriented vertically or horizontally on her nose.

◆Suggestions for further study

Find the town or city closest to the geographic center of your state by using what was demonstrated in this activity. Get a large road map of your state and trace the outline of the state on a piece of poster board. Cut out the state from the board and hang the state from one of at least three holes punched around the outer edge of the board. Use a weight on a piece of string to draw a line straight down from the hole. Do this from the other punched holes. Where the lines cross will be the geographic center of the state. Realign the actual map and the cutout map to find the city or town closest to the geographic center of the state.

◆Answers

8. The lines cross at a single point, the center of gravity.

10. Adding paper clips changes the center of gravity. The lines still cross at a single point, but not the same point as before the paper clips were added.

11. The center of gravity will always be below any point from which you suspend Cynthia. This allows you to predict which way she will be facing.

ACTIVITY 28: DEMONSTRATION

Center of Gravity: Magical Motivational Demonstrations

Materials

You will need

• a cylinder (a large holiday cookie container with lid)

• a large weight (several 2-oz fishing sinkers)

• masking tape

• an inclined plane (a board propped up on books)

• a sharp pencil

• 40 cm of fairly stiff wire or a wire coat hanger

• a lump of clay about the size of a golf ball

• an empty cereal box

• 2 styrofoam cups and lids

• 1 cup of salt (or dry, fine sand)

◆Background

Demonstrations that really grab your students' attention are good ways to motivate independent learning. This activity consists of instructions for building and demonstrating three devices that seemingly defy the laws of gravity.

◆Procedure

Part I   Uphill roller

1. Before class, tape the lead weights together on the inside wall of the cookie container. Place a small piece of masking tape on the outside of the container directly behind the weights to serve as a marker. Replace the top of the cookie container so that no one can see the weights.

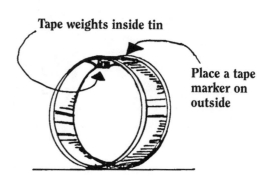

Tape weights inside tin

Place a tape marker on outside

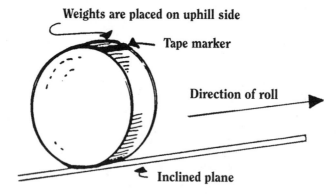

Weights are placed on uphill side

Tape marker

Direction of roll

Inclined plane

2. Place the container on the middle of the inclined plane, so that the hidden weights are slightly on the uphill side of center. (Use the masking tape marker to judge where the weights are located inside the tin.) The container will roll uphill when you release it.

3. One definition for center of gravity describes it as the single point at which the force of gravity may be considered to act. The location of the

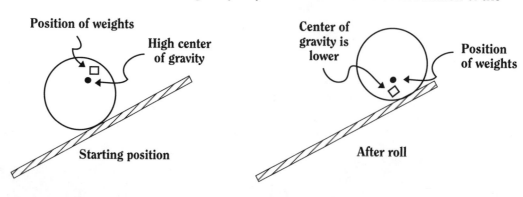

Position of weights

High center of gravity

Starting position

Center of gravity is lower

Position of weights

After roll

lead weights gives the container a very high center of gravity. When it is released, the container rolls uphill because rotating in that direction moves the lead weights downward, *lowering its center of gravity*. As the definition states, the force of gravity may be considered to act only at the center of gravity. The container moves uphill, but its center of gravity, consistent with the law of gravity, is pulled downward.

Part II   The balancing pencil

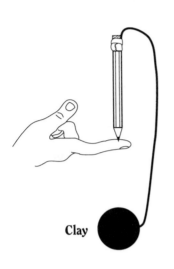

Clay

1. Challenge your class to discover a way to balance a pencil on its point on the end of a finger for 1 minute. The pencil point may not be flattened, and students are not allowed to glue or tape the pencil to the finger.

2. Demonstrate a solution by bending the 40-cm length of wire (or the coat hanger) into a C-shape.

Tape one end of the wire to the eraser end of the pencil, and place the golf-ball sized piece of clay at the other end of the wire.

3. The pencil seemingly defies gravity and "magically" balances in an upright position on your finger. The ball of clay shifts the center of gravity of the contraption to a point below the pencil point. With the center of gravity below the pencil point, the pencil will balance on its point in a nearly upright position.

Part III   Super cereal

1. Before class, tape the bottoms of the two styrofoam cups together. Punch a pencil hole through both bottoms so that salt can flow from one cup to the other. An off-center hole works best.

Snap the lid on one cup and fill the other cup about three-quarters full of salt. Now place the lid on the top cup.

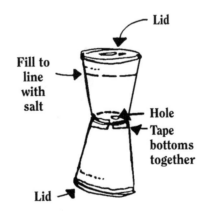

Lid

Fill to line with salt

Hole

Tape bottoms together

Lid

Tape the two cups at an angle inside the cereal box.

Turn the box upside down until all the salt is back in the top cup. Place the box on its side so that the salt stays in the top cup.

2. Tell your students that the "Super Cereal" box lying on its side contains more energy than is shown by its caloric content. To prove your statement, place the box upright on a table with half of the box sticking out over the table's edge. Suddenly, without any encouragement from you, the cereal box will tip off the table.

3. Challenge your class to develop a model of what's inside the box that would account for its strange behavior. For an extra credit home project, students may try to build a box that will behave in the same way.

Note: *Do not overfill the cup. If it does not tip the box, try removing salt.*

4. For the box to tip over suddenly, its center of gravity must move from a point *above* the table to a point that is *off the edge of the table*. This is exactly what happens as the salt flows from the upper cup to the lower cup.

Ideally, the box should remain motionless for at least 10 s, then tip over. (Longer times seem even more mysterious.) You may have to adjust the angle of the cups or the size of the hole to get it to fall on cue after a delay of about 10 s.

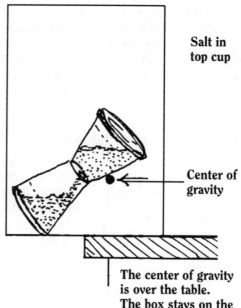

Salt in top cup

Center of gravity

The center of gravity is over the table. The box stays on the table.

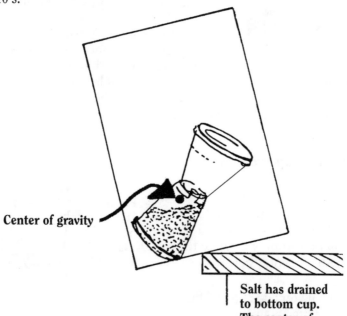

Center of gravity

Salt has drained to bottom cup. The center of gravity shifts off the table. The box tips.

ACTIVITY 29 WORKSHEET

Motorcycle Michael and Balancing Belinda's Circus Act

◆Background

Do circus high-wire performers have skills and strengths that the rest of us don't, enabling them to carry out amazing feats of balance? Maybe they do, but they also know how to use the concept of center of gravity to make their performances a lot easier than they seem.

◆Objective

To explore how applying the concept of center of gravity affects the difficulty of a seemingly impossible circus act

◆Procedure

1. Cut out the Motorcycle Michael and Balancing Belinda figures (from the patterns on the following pages) where indicated.

2. Fold the Motorcycle Michael cutout along the line near his helmet so that the bottom edges of the cutout are even with each other. Staple the open end of the cutout together along the line marked *Staple here*. (You only need three or four staples evenly spaced underneath the motorcycle wheels and seat.)

3. Carefully bend apart the bottom of the Motorcycle Michael cutout at the staples to form an inverted *V* (see diagram). We will consider this open inverted *V* to be the grooves in Michael's tires.

Rope

Staples

Inverted V

4. Tie (or tape) the string to the backs of two chairs, making a tightrope. Make sure the tightrope is as taut as possible and as far above the floor as possible.

5. Place Michael and his motorcycle on the tightrope. You will find that he is hopelessly unbalanced. Michael would indeed have to develop considerable skill to keep the cycle on the rope. If Michael is unable to balance alone, will he be able to balance when carrying Belinda?

6. Make a trapeze for Belinda by joining the three straws end-to-end. (Hint: Slit both ends of the middle straw lengthwise about 1 cm and put the cut ends inside the other straws. Secure the joints with tape.)

Materials

Each group will need
• Motorcycle Michael and Balancing Belinda pattern
• scissors
• a stapler
• 2-m length of string
• 2 chairs
• 3 average-size plastic drinking straws with flexible elbows
• 20-cm length of masking tape
• **optional:** several standard paperclips

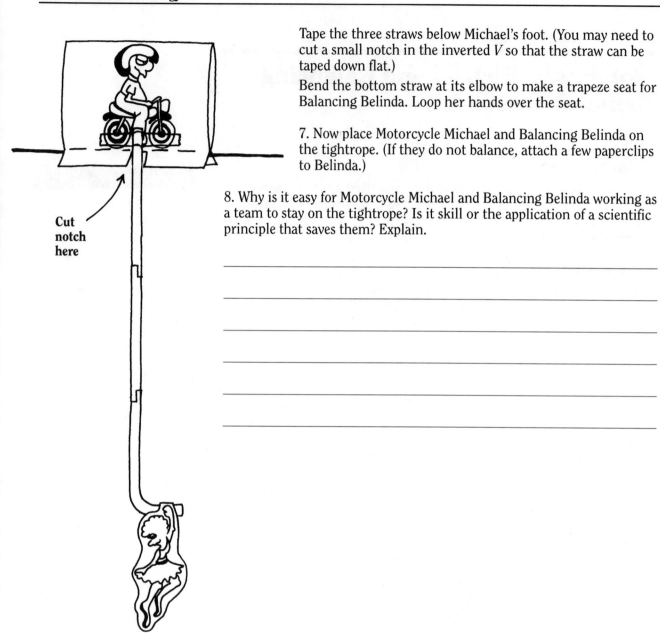

Cut notch here

Tape the three straws below Michael's foot. (You may need to cut a small notch in the inverted *V* so that the straw can be taped down flat.)

Bend the bottom straw at its elbow to make a trapeze seat for Balancing Belinda. Loop her hands over the seat.

7. Now place Motorcycle Michael and Balancing Belinda on the tightrope. (If they do not balance, attach a few paperclips to Belinda.)

8. Why is it easy for Motorcycle Michael and Balancing Belinda working as a team to stay on the tightrope? Is it skill or the application of a scientific principle that saves them? Explain.

_____

_____

_____

_____

_____

_____

Fold ⟶

Cut along
dashed line

↖ Staple here

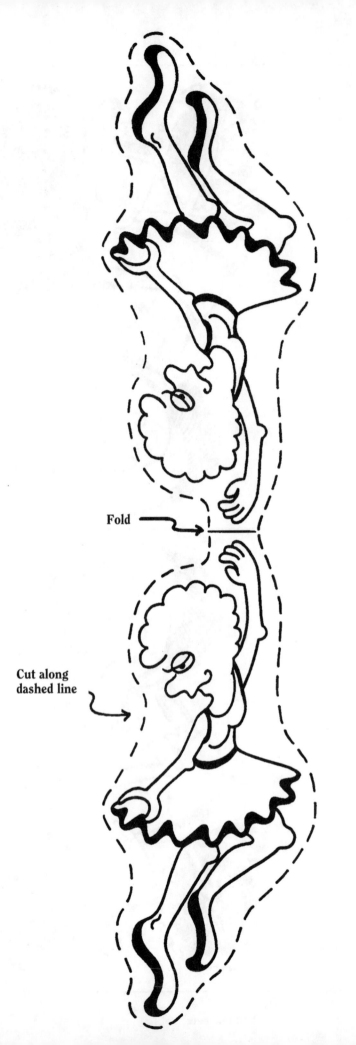

Fold →

Cut along
dashed line

GUIDE TO ACTIVITY 29

Motorcycle Michael and Balancing Belinda's Circus Act

◆What is happening?

High-wire motorcyclists frequently work with circus performers suspended below them on a trapeze that hangs from the motorcycle. Rather than risking more lives, the trapeze artists actually serve to lessen

the risk by lowering the center of gravity to a point below the wire.

When Michael and his cycle are placed on the string, they fall off. This shows that the center of gravity must be *above* the line of support. Recall that an object that is free to rotate will rotate until its center of gravity lies *directly below* the point or line of support.

To save Michael from falling, we must somehow get his center of gravity *below* the row of staples (the line of support).

The straws accomplish this. Attaching Belinda's trapeze lowers the center of gravity to a point *below* the staples. You can confirm the position of the center of gravity by finding the balance point along the straw.

◆Time management

One class period (40–60 minutes) should be enough time to complete the activity and discuss the results.

◆Preparation

Before class, make a photocopy of the Motorcycle Michael and Balancing Belinda patterns on pages 139 and 140 for each student group. If you prefer to set this up before class as a demonstration, mount Michael and Belinda on pieces of thin cardboard or on large note cards. The cardboard backing increases their durability.

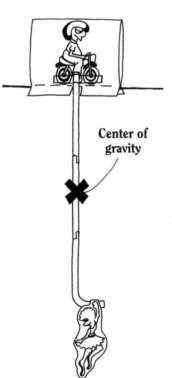

◆Suggestions for further study

How short can you make the straws and Balancing Belinda and still get Motorcycle Michael to balance nearly upright on the string? Make a tall Motorcycle Michael (half as wide and twice as tall as the Motorcycle Michael used in the activity). What is the shortest set of Balancing Belinda and straws that will keep the tall Motorcycle Michael nearly upright on the string? When Motorcycle Michael is taller, must the straws and Balancing Belinda be longer or shorter?

◆Answer

8. When Belinda joins Michael on the motorcycle, her weight and that of the straws lower the center of gravity. The heavier she is, the lower the center of gravity will be, and the harder the motorcycle will be to tip over. The lower she is suspended, the more torque she will exert. Belinda exerts more torque when suspended lower because lowering her places her at a greater distance from the string, which is the motorcycle's axis of rotation (Torque = Force x distance). The torque she exerts acts to center the motorcycle on top of the string. The more torque she exerts, the harder it is to tilt the motorcycle over sideways.

ACTIVITY 30 WORKSHEET

Clip and Trim

◆Background

Ships at sea often have to adjust the placement of their cargo or passengers in order to keep the ship level in the water; this process is called *trimming*. Knowing how to apply the scientific concepts of torque and center of gravity is essential when trimming. In this activity, you will use your knowledge of these concepts to perform a *clip and trim*—to trim a piece of cardboard by adjusting the placement of paper clips along its edge.

◆Objective

To investigate how the concepts of torque and center of gravity can be applied to keep an object balanced

◆Procedure

1. Find the middle of the strip by measuring 14 cm from one end. Mark the middle *0*.

2. Beginning at the middle, measure toward each end and place a mark along the edge every *2 cm*. Number these marks from the middle out to the end on each side. The number indicates the distance that each mark is from the middle. (You do not need to translate these distance units into centimeters.)

3. Next, use the sharp point of the scissors to poke small holes along the lower edge of the cardboard at each mark. You will hang paper clips through these holes.

Materials

Each group will need

• one rectangular piece of lightweight cardboard measuring 4 cm x 28 cm. (The piece can be cut from the cardboard back of a tablet or from posterboard.)

• a metric ruler

• a pair of scissors (or some object with a sharp point for making holes in the cardboard)

• one piece of masking tape

• 12 (or more) paper clips of the same size

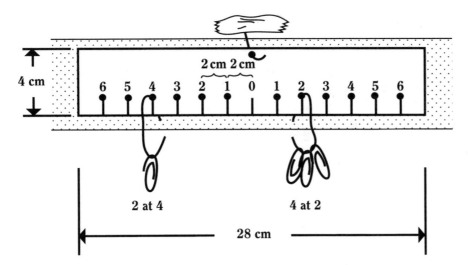

4. Poke a hole in the middle of the upper edge. This will be the balance point. Straighten out one end of a paper clip and tape it to the desk so that the straight part sticks out over the edge. Hang the strip from the paper clip.

If the strip does not balance, attach a paper clip at *0* and slide it slightly one way or the other until the strip balances. *Do not move this centering clip during the experiment and do not count it.*

5. To add weight at any position, open a paper clip so that you can insert one end into the hole in the cardboard and hang additional paper clips from the other end. When perfoming this activity, remember to count *all clips* at a given position (including the open clip attached to the cardboard).

We'll describe the clips and their positions in the following manner: If 2 clips are at position 4 on the left and they are balanced by 4 clips at position 2 on the right, we'll describe the situation as: *2 at 4 balances 4 at 2.*

6. The problem is to discover a method of accurately *predicting* how to balance the strip using paper clips.

For example, if there were 3 clips at position 5 on the left, where would you place 4 clips on the right to get the strip trimmed or balanced? The clips do not have to be hung together.

When you have discovered this method, you will have learned an important and useful concept.

In order to discover a method it is first necessary to *describe* a number of different balanced conditions. By trying different combinations, discover and record at least six different ways of balancing 3 clips at 4 on the left.

Data table

Balance conditions

3 at 4 balances _____

3 at 4 balances _____

3 at 4 balances _____

3 at 4 balances _____

3 at 4 balances _____

3 at 4 balances _____

7. Have you discovered the method or pattern? If so, put your method to the test by *predicting* where you should hang clips to obtain balanced strips for the following arrangements:

(a) _____ at 2 balances 2 at 4.

(b) 2 at 5 balances 5 at _____.

(c) 1 at 4 balances 4 at _____.

(d) 3 at _____ balances _____ at 6.

(e) 1 at 5 balances 2 at 2 and 1 at _____.

(f) 3 at 5 balances 2 at _____ and 1 at _____.

(g) 3 at 2 and 2 at 5 balances 2 at 6 and 2 at _____.

Alternatively, if you are not sure of the solution to the problem, use the challenges above to collect more information about balancing.

8. Describe a general method for predicting when an arrangement will be balanced. Do not use specific numbers in describing your method.

The strip will balance when:

_____

_____

_____

_____

_____

_____

_____

_____

_____

_____

GUIDE TO ACTIVITY 30

Clip and Trim

◆What is happening?

Combining the concepts of *torque* and *center of gravity* will help students discover a rule for balancing the strip of cardboard used in this activity. Each paper clip that is hung on the strip exerts a torque that is proportional to its distance from the axis of rotation (the midpoint of the strip, labeled *0*). The center of gravity will be located at the axis of rotation when the torques acting on both sides of the strip are equal.

There are several ways to balance the strip used for this activity. You can

• increase the *number* of paper clips on the higher arm,

• increase the *distance* between the zero point and the clips hanging on the higher arm, or

• increase both the number of clips on the higher arm and the distance between those clips and the zero point.

After experimenting with these techniques, students may discover a method similar to the following for *predicting* how to balance the strip:
• At each position where clips are hung, multiply the number of clips by their distance from the center.

• Add these products together for each side of the strip.

• The strip balances when the sum of the products for one side of the strip is equal to the sum of the products on the other side.

Let's examine some calculations showing how this works for a specific case: Does 1 clip at position 4 balance 2 clips at position 2? Following the steps given above, you can predict whether the strip will be balanced by multiplying 1 x 4 and 2 x 2. Since both products equal 4, the strip is balanced.

This method also works when several clips are placed at different distances from the center on one side. In this case, you simply have more than one product for a side. For example: Does 1 at 5 balance 2 at 2 and 1 at 1? We have the product of 1 and 5 for the left side, and *two* products for the right: the product of 2 and 2 and the product of 1 and 1. Calculating the sums of the products shows that the strip is balanced in this situation, because 1 x 5 = (2 x 2) + (1 x 1).

What causes the strip to balance is not the number of clips, but the *weight* of those clips. Since each paper clip used for this activity weighs about the same, we can use numbers of clips to predict when the strip will be balanced. However, to be more precise, we should multiply the *distance* times the *weights* of the clips.

Weight is a force. What we are really doing is multiplying force times distance. Torque is defined as the product of force and the distance from the rotation point to the point of force application. So, the general method for balancing the strip can be stated in terms of torque: *The strip will be balanced when the sum of the torques on one side equals the sum of the torques on the other side.*

◆Time management

One class period (40–60 minutes) should be enough time to complete the activity and discuss the results.

◆Preparation

Encourage individual students to experiment with the apparatus, but do not be surprised if a large percentage of your students have difficulty applying the concept of torque to balancing tasks. Discovering the exact rule for balancing the strip is a difficult challenge that requires high-order reasoning ability.

◆Suggestions for further study

The principles explored in this activity also can be used to explain the way a two-pan balance works. Help students apply the concepts of torque and center of gravity to the function of this common laboratory instrument.

◆Answers

6. Three clips placed at position 4 on the left arm balances the following combinations of clips placed on the right arm (these are not the only possibilities):

3 at 4

4 at 3

2 at 6

6 at 2

2 at 2 and 2 at 4

2 at 2 and 4 at 2

3 at 2 and 2 at 3

3 at 1 and 3 at 3

1 at 3 and 2 at 2 and 1 at 5

2 at 1 and 1 at 2 and 2 at 4

1 at 1 and 2 at 2 and 1 at 3 and 1 at 4

7. Combinations of clips that balance:

(a) 4 at 2 balances 2 at 4.

(b) 2 at 5 balances 5 at 2.

(c) 1 at 4 balances 4 at 1.

(d) 3 at 4 balances 2 at 6 or 3 at 2 balances 1 at 6 or 3 at 6 balances 3 at 6.

(e) 1 at 5 balances 2 at 2 and 1 at 1.

(f) 3 at 5 balances 2 at 6 and 1 at 3.

(g) 3 at 2 and 2 at 5 balances 2 at 6 and 2 at 2.

8. The strip will balance when the sum of the products for one side is equal to the sum of the products for the other side. Each side can have from one to six products (corresponding to the six positions on each side); each product is the result of multiplying the position number (the distance from the center) by the number of clips hanging there.

EVIDENCE OF ENERGY

Readings

The following readings review and expand upon the concepts introduced in the activities in Modules 1–6. Many teachers may wish to learn more about a particular topic than is included in these inservice materials. *Conceptual Physics* by Paul G. Hewitt (Little, Brown, and Company, 1985) is an excellent textbook to use for this purpose.

Dr. Hewitt encourages readers to develop an intuitive understanding of the everyday applications of physics; he does not stress "number crunching." The book's illustrations are both amusing and informative. Many chapters list home projects, simple experiments, demonstrations, and tricks for illustrating a particular concept.

Reviewing Newtonian Mechanics

◆Introduction

Many of the basic terms and concepts used in these modules were introduced in *Methods of Motion*, the first volume of this two-part series. The activities there served to introduce students to mechanics, and in particular to Newton's three laws of motion. While the activities in this volume are designed to be independent of those in the first, explaining the results obtained does require the use of some topics covered in the first volume. This reading reviews these basic concepts.

◆Mechanics

Mechanics is the branch of physical science that describes the behavior of bodies in motion. Much of our knowledge of mechanics is based upon the work of Sir Isaac Newton. In 1687, Newton published a brilliant synthesis of the principles of mechanics, *Philosophiae Naturalis Principia Mathematica* (commonly known as *Principia*). Newton's insights were so powerful and complete that they remained virtually unchallenged and unmodified for two centuries.

◆Newton's laws of motion

Newton is best known for his three laws of motion. These laws explain the way that matter moves (or *does not* move).

First law: The law of inertia

An object at rest tends to stay at rest, and an object in motion tends to stay in motion in a straight line and at a constant speed unless acted upon by an unequal force.

This law emphasizes that setting an object in motion or stopping an object that is already moving both require the application of a force (a push or pull) to the object. Things *do not* begin to move or stop by themselves.

Second law: Interactions of force, mass, and acceleration

Force equals mass times acceleration ($F = ma$).

The second law gives the mathematical relationship between *force* (how hard you push or pull on an object), mass (the amount of "stuff" being pushed or pulled), and acceleration (the rate of change of speed of the object).

Third law: The law of opposing forces

This law can be stated in either of two ways:

For every force there is an equal and opposite force.

or

To every action there is an equal and opposite reaction.

The essence of this law is the idea that when you push against something (apply a force), the object "pushes back" (applies an equal and opposite force) just as hard. This points out an important aspect of forces: A force is a push or a pull in a particular *direction*. You have not completely described a force until you have specified not only its strength, but its direction too.

◆Mechanics terminology

The following terms are also included in the glossary at the end of the book.

Acceleration: The rate at which an object speeds up or slows down. By Newton's second law, in order for an object to be accelerating, a force must be acting on it. Acceleration is occurring if an object travels *different distances* during equal time intervals that its motion is observed. The units for acceleration are meters per second per second, or meters per second squared (m/s^2).

Air resistance: A force exerted on a moving object opposite to its direction of motion due to the friction between the object and air. Air resistance is also called *drag* or *air friction*.

Dynamics: A branch of mechanics that concentrates on the study of bodies in motion.

Friction: Resistance to relative motion between objects in contact. The force due to friction acts on an object in the direction opposite to that of its motion.

Force: A push or pull in a particular direction that can be applied to an object. We can more technically define a force as something that has the capacity to change the motion of an object. Both the *magnitude* (strength) and the *direction* must be stated when defining a force. Vectors are often used to represent forces.

Gravity: A phenomenon that exists throughout the universe. The force due to gravity is the force of attraction that exists between all objects in the universe. The amount of gravitational force between two bodies (such as the Earth and a rock thrown up into the air) depends on the mass of both objects and the distance between them. Earth's gravitational force is just one example of the general phenomenon of gravity. On Earth, the force due to gravity is the force that causes objects (such as an airborne rock) to accelerate toward the Earth.

Inertia: A measure of an object's resistance to change in motion. Inertia is another way of describing an object's mass. The more mass that an object possesses, the more force that is required to set it in motion or to stop it from moving. Inertia is a property possessed by all matter that can be thought of as laziness or "difficult-to-moveness."

Mass: A property of matter related to inertia. As the mass of an object increases, so does its inertia. Mass can be thought of as the quantity of matter in an object. Mass is *not* the same as weight.

Mechanics: The branch of physical science that describes the behavior of bodies in motion; mechanics deals with energy and forces and their effects on bodies.

Newton: The standard metric unit of force. One newton (1 N) is the amount of force required to accelerate a 1000-g mass at a rate of approximately 1 m/s^2.

Speed: The rate of motion; speed combines information about how far an object travels with how long it takes to travel that distance. An object's speed is calculated by dividing the distance traveled by the time interval. The metric units for speed are meters per second (m/s). Speed only tells us about the magnitude of motion, not the direction. An object moving at a *constant speed* will move the same distance during each successive time interval.

Vector: An arrow that can be used to represent quantities such as force or velocity. The *head* of a vector shows direction; the *tail* shows the magnitude or strength. A long tail indicates a strong force; a short tail indicates a weak force.

Velocity: A measure of motion that specifies both the *direction* of motion and the *magnitude* of motion. Velocity is *not* the same as speed, because the speed of an object tells how fast the object is moving, but does not specify the direction in which it is moving. Speed is actually the magnitude of velocity.

Weight: On Earth, a downward force that acts on an object due to the gravitational attraction between the object and Earth. *Weight is not the same as mass.* On the moon, you would only weigh 1/6 as much as you do on Earth, because the moon's gravitational attraction is less than Earth's. You would still have the same amount of mass, however, even though you would weigh less on a spring scale.

READING 2

Projectile Motion

◆Introduction

Projectile motion has captured our interest since ancient times. People have studied the flight of stones flung from leather slings, arrows cast at foes, boulders launched from catapults, and most recently, rockets blasted into space.

◆Intuitive knowledge of projectile motion

Each of us has learned about projectile motion from everyday experiences. When a baseball is hit or thrown in our direction, we intuitively judge the trajectory or path of the ball and quickly decide how fast and in what direction we must travel and when and where we should place our hands so ball and hands meet. Even more sophisticated knowledge of projectile motion is demonstrated by a football quarterback who must observe the speed and direction of his receiver and launch the ball with just the right force and direction so that ball and receiver reach the same point at the same time.

The instinctive nature of throwing or catching a ball demonstrates our practical knowledge of how things move. Our formal knowledge and understanding of motion often lag behind our intuitive knowledge. Many major league outfielders who almost never miss a fly ball would "strike out" if asked to demonstrate formal knowledge of projectile motion by accurately drawing the path of a pop-up or explaining the forces which combine to produce this path.

◆Historical perspective

Twenty-three hundred years ago, Aristotle stated that projectiles need some force to keep them moving through the air. He postulated that air "pinches in" behind the projectile and squeezes it forward in much the same way we might squeeze a slippery apple seed to send it shooting off. Aristotle's beliefs remained unchallenged until the 17th century when Galileo Galilei, addressing questions about moving and falling objects, developed methods of experimentation and reasoning that revolutionized science.

Galileo's challenge to Aristotle may in fact mark the beginning of modern science. Through a brilliant combination of reasoning and experimentation, Galileo convincingly argued that projectiles need not have "rear pinching air" to keep them moving. Galileo's work laid the groundwork for Sir Isaac Newton's laws of motion.

Newton's elegant synthesis proved so powerful for predicting many cases of motion that some experts declared further research into mechanics unnecessary. However, 20th century scientists found his laws inadequate for explaining motion at the subatomic or galactic scale. Just as Newton built on Galileo's work, Newton's laws have, in turn, become springboards for the 20th century's new physics of quantum mechanics and relativity.

◆A quick review of straight-line motion

Newton's first law of motion states: An object at rest tends to stay at rest, and an object in motion tends to stay in motion in a straight line and at a constant speed unless acted upon by an unequal force.

On Earth, multiple *hidden* unequal forces (including gravity and air resistance) prevent objects from moving in a straight line at a constant speed. These forces are hidden in the sense that we tend to overlook them because they are invisible or so common that we take them for granted. Despite these pervasive *unequal* forces (forces not balanced by an equal opposing force), we can use a ball to demonstrate straight-line motion in two ways:

• by rolling it along a smooth, level surface

or

• by dropping it straight down.

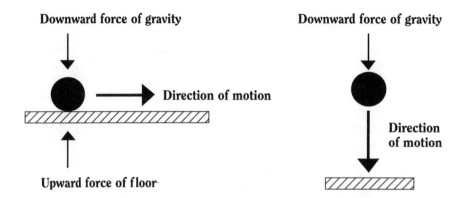

Under these sets of conditions, gravity cannot cause the ball to curve. When the ball rolls across a flat surface, the force of gravity acting on the ball is canceled out by the equal opposing force exerted on the ball by the floor. When an object is falling straight down, gravity cannot change the object's path because it is acting in the same direction as the motion of the object.

◆Projectile motion—What's the difference?

A projectile is a body set in motion by some external force. As the first law of motion states, a body in motion tends to stay in motion in a straight line and at a constant speed unless affected by an unequal force. In real situations, almost all moving objects are affected by unequal forces, so very few objects move in straight lines. The actual path that any object travels is determined by the sum of all the unequal forces acting on it. These forces may be obvious (the thrust of a rocket engine) or hidden (gravity or air resistance).

The activities in Module 1 allow you to investigate the motion of projectiles moving along a curving path. The curving motion occurs because an unequal force affecting the object is acting at an angle (other than 0° or 180°) to the direction that the object is launched.

This type of motion can be represented on a graph as a series of closely spaced points that combine to form a curving line. Each point is defined by a pair of vertical and horizontal coordinates

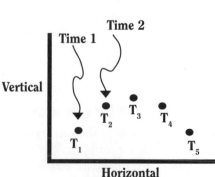

that represent the position of the object at a particular time.

Even though projectile motion occurs in a real, three-dimensional world, the cases that we will study have effectively only two dimensions—displacement in the horizontal dimension, and up or down motion in the vertical dimension.

◆Why do projectiles curve? Why do they stop curving?

The activity "Bent Barrel Ballistics" examines the motion of a projectile traveling along a curving path inside a tube. Before releasing the ball, the experimenter is asked to predict the result.

Will the ball continue to curve in the same direction after leaving the tube, curve in a different direction, or straighten out after being shot out of a curving tube?

The key to correctly predicting the experimental result is suggested by part of the first law of motion: ". . . an object in motion will continue in motion in a straight line . . . unless acted upon by an unequal force."

Now the questions become: "What makes the ball curve?" "What makes the ball *stop* curving?" The ball curves inside the tube because the sides of the tube exert an unequal force on one side of the ball. Once the ball escapes from the tube, the forces acting on it are (approximately) equal. In the absence of unequal forces, the ball travels in a straight line once it leaves the tube.

The activity "Tunnel Trajectories" requires the experimenter to approach the straight path versus curving path problem from a different standpoint. The problem is stated in such a way that many people will try to cause the ball to curve as it goes through the tunnel. They quickly discover that *if you follow the rules, you cannot cause the ball to curve after it is released*.

Once the unequal force applied by the experimenter's hand is no longer acting on it, the ball will stop curving and roll along a straight-line path as the first law of motion predicts.

In summary, both "Bent Barrel Ballistics" and "Tunnel Trajectories" demonstrate that when an unequal force producing curving motion is removed, the path of a moving object straightens out.

◆Predicting the path of projectiles

Predicting the path of an object dropped from a moving platform can be tricky. There *is* an unequal force acting on such objects: *the force due to gravity*.

The force of gravity will accelerate any object toward Earth at about 10 m/s^2. But gravity is working at right angles to the direction of the ball's motion. This means that an object will not continue moving in a straight line. It will simultaneously move *forward* (its original direction of motion) and *downward* (the direction in which the force of gravity is pulling the object). The combination of the forward and downward motion makes the object travel along a curving path after it is released.

READING 3

Dissecting Projectile Motion

◆Introduction

Newton's laws of motion are simple to state, and it is not too difficult to apply the laws to predict simple patterns of motion. However, practical experience tells us that projectile motion is rather complex. Thinking about *multiple* factors that are simultaneously affecting a moving object is tricky, but accounting for *all* the factors affecting the motion of an object traveling through the air near the surface of the Earth is the key to understanding projectile motion.

◆Predicting the path of projectiles: Examining one factor at a time

To simplify the analysis, for the following case of motion we will examine the trajectory of a projectile in an imaginary world where we have complete control over the factors which affect its motion. The example is based on an all-too-common type of projectile motion: firing the gun of a tank.

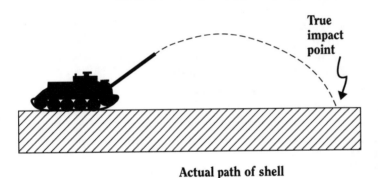

Actual path of shell

When a tank fires its cannon, the shell has a trajectory similar to the one shown in the diagram to the left. Using special equipment, it is possible to photograph the flight of a cannon shell. The diagrams that follow are similar to such photographs. (In order to emphasize the general pattern of motion, distances are not drawn strictly to scale.)

The shell's trajectory will be shown for several *imaginary* or ideal sets of conditions. These conditions are imaginary only in the sense that they do not normally exist on Earth. Methods of approximating such conditions are described for each case.

Imaginary condition 1: No gravity and no air friction

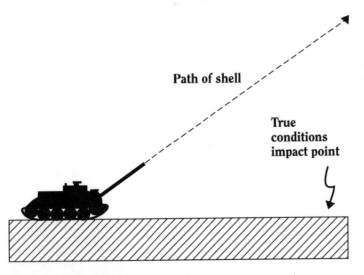

Under these conditions (similar to that found in interstellar space), the shell would exhibit the straight-line, constant-speed trajectory shown to the left. No forces act on the shell after it leaves the tank's muzzle.

The shell in this imaginary situation would never hit the ground; it would continue moving in a straight line. If the location of the projectile were noted each second after firing, the distance traveled during each 1-s interval would be equal, since the shell is traveling at a constant speed. Newton's first law of motion predicts this simple motion.

Imaginary condition 2: Gravity, but no air friction

Under these conditions (that theoretically could be produced in a huge vacuum chamber on Earth), the shell would not be slowed by friction between the shell and the air. It would curve and fall back to Earth as a result of gravitational force, but the *horizontal* component of the shell's motion would remain the same throughout its flight.

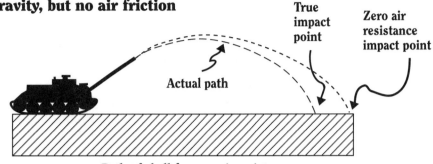

True impact point

Zero air resistance impact point

Actual path

Path of shell for zero air resistance

Under actual conditions on Earth, air resistance slows cannon shells by a substantial amount. The distance between the actual impact point and the one shown in this ideal condition is proportional to the force of air resistance. (Numerical values for air resistance are the product of surface area multiplied by the square of the speed. Since the speed is changing during the shell's flight, a detailed analysis of the actual air resistance at any point is a complicated problem and beyond the scope of these materials. For this reason, the activities in the workshops are designed to minimize the effects of air resistance.)

◆Is gravitational acceleration affected by horizontal motion?

Combining the paths for the two imaginary conditions allows us to answer this question: *No!* Although it may seem intuitively reasonable to assume that something moving forward at high speed should be affected less by gravity and fall downward more slowly, this is *not* correct! *Downward acceleration due to gravity is unaffected by horizontal motion. Deviations from ideal patterns of motion are caused by hidden forces such as air resistance.*

◆Summing it up: The effects of gravity and air resistance on a high-speed projectile under realistic conditions

The actual path of a projectile is determined by the sum of all the forces acting upon it. If all the hidden forces acting on it are accounted for, the shell will be found to be moving in the path predicted by Newton's laws of motion.

The force of air resistance pushes against any shell moving through Earth's atmosphere and slows it down. Since real tank shells travel at very high speeds, the air resistance has an effect that is large enough to affect the path of the shell by a significant amount.

The following diagram shows the actual path of the shell compared with the trajectories we plotted when we "turned off" gravity or air resistance in the imaginary conditions 1 and 2.

Ideal path: Zero gravity and zero air resistance

Zero air resistance path

Effect of gravity

Actual path

Effect of air resistance

READING 4

Psyching Out the Psycho Cyclist

◆Introduction

The tank described in Reading 3 fired a projectile in an ideal environment in which the effect of each force acting on the projectile could be isolated and examined individually. Unfortunately for physics students, in our real environment on Earth all of the forces affecting the path of a projectile operate simultaneously. The following examples show how real projectiles move near the surface of the Earth.

◆Predicting the path of Psycho's bowling balls

Predicting the path of an object dropped from a moving platform is tricky, because two components of motion must be accounted for: the downward acceleration of the ball due to the force of gravity acting on the ball, and the forward (horizontal) motion parallel to the path of the moving platform. Let's examine the downward and forward components of the motion of a bowling ball dropped by a Psycho Cyclist.

Speeding Psycho is here when he drops the ball

Speeding Psycho is here when the ball hits the ground

A   B   C   D

The ball's downward motion is caused by gravity. The force due to gravity is an unequal force in this case, accelerating the ball toward Earth at about 10 m/s^2. The ball moves downward when Psycho releases it because there is no longer an upward force exerted by his hand equal to the downward force of gravity. But gravity is working at right angles to the direction of the ball's horizontal motion. This means that the ball will *not* fall straight down. It will follow a curving path as it falls.

Now consider the horizontal (forward) component of the ball's motion. Once Psycho lets go of the ball, he cannot apply any additional horizontal force to it. In this situation, the first law of motion predicts that the ball will continue moving forward at a constant speed unless some force acts on it in the opposite direction. Only the force of air resistance is acting to slow down the ball's forward motion, and at the slow speed that Psycho is riding his bike, the effect of this force is almost zero. Therefore, it is reasonable to assume that air resistance will have no observable effect on the forward speed of the bowling ball. As it falls downward, the horizontal component of the ball's motion will remain equal to the speed of the bike.

Of course, in real life the horizontal and vertical motions of the ball occur simultaneously. We cannot isolate each of the two components of motion. We observe the ball traveling downward along a curving path because of its combined downward motion and horizontal (forward) motion.

Now that you have thought some more about the motion of the ball and the forces affecting its motion, look again at the illustration of the Psycho Cyclist. Where do you think the ball will land?

If you do not remember how projectiles behave under ideal conditions, some puzzling questions may occur to you: Will the horizontal motion of the ball in some way cancel out or add to the downward acceleration caused by gravity? Will gravity make the ball move faster or slower horizontally? Will it fall to Earth faster or slower as a result of its forward motion? The answer to all of these questions is: *No! The gravitational acceleration and horizontal speed are independent of one another!*

Let's take another look at the real conditions as Psycho drops the ball.

The center of the ball and the center of the rear wheel of the bike are both directly above point A when the ball is released. After the ball is released, no force is pushing the ball forward and so it continues moving in the same direction as the bicycle because of inertia (Newton's first law). When it is released, the ball is traveling at exactly the same speed as the bicycle. Since the friction between the ball and the air is (assumed to be) negligible, the bowling ball continues moving *forward* at the same speed as the bicycle until it strikes the ground.

It begins to accelerate

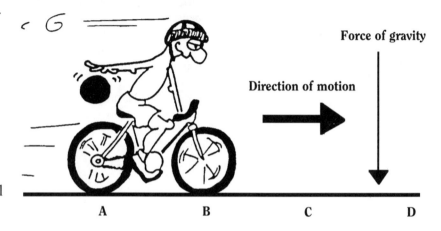

downward because of the unbalanced force of gravity pulling it toward the Earth with an acceleration of about 10 m/s². It continues its forward motion while it falls under the influence of gravity. This combination of forward and downward motion produces a curving path.

Since the ball and the cycle are *both* moving *forward* at exactly the same speed, the ball keeps up with the bike as it falls. This means that as the ball falls, it will remain aligned with the center of the bike's rear wheel. The vertical fall of the ball does not affect the forward component of its motion. When it hits the ground, the ball will still be even with the center of the wheel. Therefore, the correct answer is D.

◆How can a falling projectile move forward at a constant speed?

One of the main impediments to understanding projectile motion is the true but counter-intuitive statement that horizontal motion and vertical acceleration due to gravity are independent of one another. Gravity does not cancel out forward motion, and forward motion does not slow down gravitational acceleration. The following examples illustrate some of the practical implications of this proposition.

Imagine traveling in a commercial airplane. Once you and the plane are up and traveling at a constant speed of 600 km/hr, *the forward motion of your body does not depend on the airplane.* Your continuous straight-line, constant-speed motion is predicted by the first law of motion.

Now suppose that in some *imaginary* airplane, the first law of motion does not apply once the plane leaves the ground. Flying in this airplane would be dangerous! If the first law of motion is repealed, your forward motion can only continue as long as the plane is pushing you. If you somehow lost contact with the plane (for example, by making a small jump while walking inside the cabin), your forward motion would suddenly stop. The back wall of the cabin of the plane would still be moving forward at 600 km/hr and would smash into you.

Yet another analogy for a projectile's forward motion can be based on your own bicycle-riding experience. You can think of the projectile as coasting forward after it is set in motion—just like you coast on a bike.

It takes a force to accelerate a bike up to top speed (F = ma; Newton's second law), but if you stop pedaling while at top speed you do not suddenly stop. There is no longer a force (from pedaling) moving you forward, but you keep moving forward because of inertia (Newton's first law). The force of air resistance and friction acting on the wheels will eventually slow you to a stop, but this takes a while. Similarly, a projectile such as a cannonball keeps moving forward at a constant speed after it leaves the muzzle. The downward force of gravity does not affect its forward speed.

On the basis of your knowledge of projectile motion, try to predict the outcome of the following simple experiment: You are walking at a fast, constant rate across the room, and while doing so you toss your keys straight up in the air. Where will the keys land: behind you, in front of you, or back in your hands? Do the experiment and see!

READING 5

How Hard Are You Working?
What Are You Working Against?

◆Introduction

Using the formula *Work = Force x distance* to calculate the amount of work being done seems very simple. However, this formula cannot be applied to a moving object without considering *all* the forces acting on the object. All matter in the universe is constantly affected by hidden forces that may change the direction and/or speed of a moving object in ways that can lead an unwary observer to an incorrect conclusion about how much work is being done.

◆Work and the direction of force

In order to describe a force, you must know two pieces of information: the *magnitude* (strength) of the force and the *direction* of the force.

Knowing the direction of the force is also essential for calculating work. Work is done by a force only when the object is moving in exactly the *same direction* or the *opposite direction* as the force. Let's analyze the work being done in the following example.

Apple Juice (AJ for short), an ex-football player making a commercial for a rental car company, picks up a heavy suitcase and dashes through a crowded airport. If the suitcase weighs 200 N and he lifts it to a height of 0.1 m, how much work does he do by lifting it?

When AJ lifts the bag straight up, the formula Work = Force x distance can be applied in a direct fashion. For this case,

Work = (weight of the bag) x 0.1 m = 200 N x 0.1 m = 20 J.

Simply multiplying the weight of the suitcase (200 N, a downward force) by the distance that Juice lifts the suitcase (0.1 m) gives the correct result:

20 N • m, or 20 J of work.

Experimentally determining the force being applied to the suitcase *while it is being lifted* presents some problems, even for simple examples such as this. If you attach a spring scale (calibrated in newtons) to the suitcase, you will obtain at least *three different force readings* as the suitcase is lifted from the floor to a height of 0.1 m:

• As the 200-N bag accelerates off the floor, the scale will show that a force *greater* than 200 N is being exerted.

• While it is moving upward at a constant speed, the scale will read 200 N.

• When the bag decelerates as it approaches the height of 0.1 m, the scale will show a force of *less* than 200 N.

• When the bag comes to rest at a height of 0.1 m, the force reading on the scale will return to 200 N, the bag's weight.

Using a value other than the bag's weight (200 N) to calculate the work required to lift it 0.1 m gives an incorrect result. (Note: You can use a weight attached to a small spring scale to demonstrate these changing readings.)

◆Determining work for objects that are moving horizontally

When an object is moving at right angles to the force of gravity, the weight of the object is *not* the force that is used to calculate the work being done. For an object being moved along a horizontal surface, the force that determines how much work is required to maintain a constant speed is the *force due to friction*.

Let's look at what AJ does after the filming is over. He has had to run so far carrying the suitcase that he is exhausted. AJ decides to drag it. Walking at a constant speed of 2 m/s, he uses a spring scale to check how much force he is applying to the suitcase. He finds that the force reading is 50 N—much less than the weight of the case.

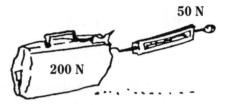

Dragging the bag at a constant speed

Since AJ is moving the bag horizontally, the force of friction between the bag and the floor determines how much force he must apply to keep it moving at a constant speed. (Newton's second law, $F = ma$, shows that if the net force acting on an object is zero, then the acceleration of the object is zero. So if AJ applies a force to the bag that is equal and opposite to the force of friction, the horizontal net force is zero and the bag does not accelerate; i.e., it moves at a constant speed.) Therefore, to calculate the work required to move the bag *horizontally* a distance of 0.1 m, you substitute the force required to drag the bag at a constant speed in the equation for work. The force required to drag the bag is 50 N, which is numerically equal to the force of friction acting on the bag. In this case,

Work = (force required to drag the bag) x 0.1 m = 50 N x 0.1 m = 5 J.

The work required to drag the 200-N bag 0.1 m may not always be 5 J, however. AJ discovers that when he drags the suitcase across a slick, freshly-waxed area of floor, the friction between the bag and the floor is reduced to 25 N. On this slick (low-friction) surface, moving the 200-N bag a distance of 0.1 m only requires 2.5 J of work (25 N x 0.1 m = 2.5 J).

The work required to drag the suitcase also changes when AJ allows one of his many fans (a child weighing 200 N) to sit on it while he drags it along. Juice notices that when he is dragging the suitcase horizontally, increasing its weight by 200 N *does not* increase the force required to drag it by 200 N.

Dragging the bag plus a fan

200 N

200 N

The increased friction between the bag and the floor is not equal to the added weight of the passenger, but it did increase proportionally with the weight addition. Juice finds that he has to exert only 100 N of force to keep the 200-N bag plus the 200-N passenger moving at a constant speed of 2 m/s. He must do 10 J of work for every 0.1 m that he *drags* the suitcase and passenger (100 N x 0.1 m = 10 J). *Lifting* the suitcase and passenger 0.1 m would require 40 J of work (400 N x 0.1 m = 40 J).

◆Can *not* working be tiring?

AJ finds that even dragging the suitcase is more work than he wishes to do. He unloads the fan, picks up the bag, and steps onto the airport's moving sidewalk.

AJ and the bag are whisked along at a constant speed of 5 m/s. However, his arm and shoulder soon start to feel tired from holding the bag 0.1 m above the moving sidewalk. Is he getting tired from the work that he is doing moving the bag? How much work *is* AJ doing by holding the bag as the sidewalk moves?

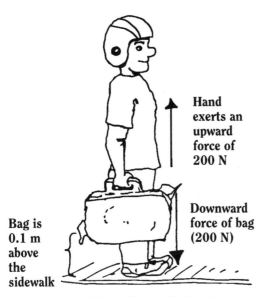

Hand exerts an upward force of 200 N

Downward force of bag (200 N)

Bag is 0.1 m above the sidewalk

Sidewalk speed is 5 m/s

Let's identify the forces involved. When AJ is holding the suitcase 0.1 m above the moving sidewalk, his hand exerts an *upward* force of 200 N on the bag. The force of his hand is opposed by a 200-N *downward* force exerted by gravity on the bag. Since the upward force of his hand on the bag exactly balances the downward force of gravity on the bag, the suitcase stays 0.1 m above the ground as it moves forward. The bag is not moving any vertical distance, so *zero work* is being done in the vertical direction.

Is he doing any work in the horizontal direction? Both AJ and the suitcase are moving horizontally along the moving sidewalk at the sidewalk's constant speed of 5 m/s. Again, let's examine the forces acting on the bag.

The horizontal force of friction opposing the suitcase's forward motion is extremely small—essentially zero. According to the first law of motion, *an object in motion tends to stay in motion in a straight line and at a constant speed unless acted upon by an unequal force.* Since there are no unequal forces acting on it, inertia accounts for the forward motion of the suitcase.

Juice does not have to apply any forward force to the bag to move it 5 m horizontally each second. The 200-N force that AJ applies to the suitcase is

Zero forward force

Friction from air is approximately zero

upward—*not in the direction that the case is moving*. Since he does *not* supply any force in the direction of the suitcase's motion (horizontal), he is doing *zero work* on the suitcase.

AJ's arm really is getting tired, but not because of the work it is doing. The suitcase is exerting a *torque* on his shoulder joint. You will study torque in Module 5.

After reviewing the film, the director of the commercial calls AJ back to the airport for a set of retakes. During the filming, a practical joker switches suitcases, substituting a suitcase that contains a set of barbells weighing 2000 N. AJ (who is a bit out of shape now that he is retired) strains and sweats, but cannot pick up the 2000-N suitcase. How much work does he perform during this futile attempt?

Despite his struggles, AJ does zero work. The large force he applies does not move the suitcase even one millimeter in the direction of the force. A large force multiplied by zero distance gives zero work.

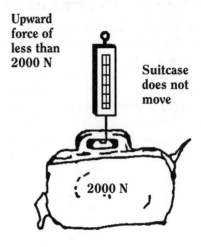

Upward force of less than 2000 N

Suitcase does not move

2000 N

READING 6

Forms of Energy

◆Introduction

Energy exists in many different forms. This reading describes common forms of energy and briefly discusses energy transformation. In principle, any form of energy can be changed into any other form. Maintaining industrial societies requires energy transformations on a grand scale. Life itself depends on the ability to transform energy.

◆Work and energy

In these modules, energy is defined simply as "the ability to do work." Any time that a force is moving through a distance (the definition of work), energy is present.

To simplify the task of describing energy, we divide it into two general categories:

• **kinetic energy**, the energy due to an object's motion

• **potential energy**, the energy due to an object's position or the arrangement of its parts

Kinetic and potential energy can be described using mathematical formulas. The kinetic energy of a moving object is equal to one-half its mass times its velocity squared (or $KE = 1/2mv^2$). Gravitational potential energy, the energy due to an object's position in a gravity field, is equal to an object's weight times its height above some reference plane. In more general terms, it is equal to the object's mass times the constant g (the acceleration due to gravity on Earth, approximately equal to 10 m/s^2) times the height (usually written $GPE = mgh$). Other forms of energy can be represented using other formulas.

In these modules the emphasis is placed on *identifying* the types of energy present and recognizing the *energy transformations* that are taking place, rather than on calculating the amount of energy that is present. Since the terms *potential* and *kinetic* energy are very general, having smaller subcategories of types of energy makes the bookkeeping on energy transformations easier. The following section defines these subcategories and discusses each of them briefly.

◆Types of kinetic energy: Thermal energy, light, sound, and electricity

The atoms and molecules of all substances are constantly vibrating and moving in a haphazard fashion. Physicists call this type of molecular motion *thermal motion*. Thermal motion cannot be seen directly because of the small size of molecules. **Thermal energy** is the energy an object has as a result of the thermal motion of its atoms and molecules.

Physicists avoid using the terms *heat* or *heat energy* as synonyms for thermal energy. Physicists define **heat** as the thermal energy transferred from one object to another as a result of the temperature difference between the two objects. Heat should *not* be thought of as being contained in an object; heat is defined only in terms of a *process of energy transfer* between two objects having different temperatures.

There are several commonly used units for describing the quantity of heat: the calorie, the Calorie, and the BTU. A **calorie** is the amount of heat required to change the temperature of 1 g of water by 1° C. The **Calorie** (or **kilocalorie**) is commonly used in rating foods; it is equal to 1000 calories. The heating ability of furnaces is often described in terms of British thermal units (BTU). A **BTU** is the amount of heat required to change the temperature of one pound of water by 1° Fahrenheit. Notice that the calorie, Calorie, and the BTU all define the quantity of heat in terms of a *change* that accompanies the process of transferring energy from one object to another.

Heat can flow only from a hotter object to a colder object. In order to predict whether or not heat will flow from one object to another, you need a way of measuring the "hotness" of both objects. **Temperature** is a quantitative way of stating "how hot" an object is. Temperature indicates the average kinetic energy of the atoms and molecules of an object. For scientific purposes, temperature is measured in degrees Celsius.

Temperature can be used to predict the direction of heat flow between two objects. Heat does not necessarily flow from an object possessing more thermal energy into an object having less thermal energy. For example, there is more thermal energy in a bathtub of warm water than in a red-hot sewing needle, but if the red-hot needle is dropped into the water heat will not flow from the warm water to the needle. Instead, heat will flow from the hot needle to the bath water. The bath water has more thermal energy than the needle, but its temperature (the *average* kinetic energy of its atoms and molecules) is lower.

Heat flows between objects every time energy is transformed; in many energy transformations, unwanted heating is a major "energy thief." For example, a "perfect" light bulb would convert 100% of the electrical energy that it receives into light energy. In fact, less than 10% of the electrical energy passing through incandescent bulbs is converted to light. Most of the energy heats the bulb and its surroundings.

Any time two moving surfaces come into contact in a machine, heat flows between the parts. Using lubricants, ball bearings, and other friction-reducing devices helps to minimize this heating. Automobiles and some other machines require cooling systems to prevent damage to their moving parts caused by overheating. Cooling systems are designed to lower the temperatures of moving parts by allowing their "waste" thermal energy to heat a liquid or gas coolant.

Sound energy is a wave phenomenon; all sounds consist of waves produced by the vibrations of the molecules of an object. Transmitting sound energy requires the presence of matter. The denser the material, the faster sound travels through it. Sound energy cannot travel through a vacuum.

Students often find it hard to believe that sound can really do work. How can you convince them that sound is truly a form of energy and that it can apply a force through a distance? Some dramatic but rarely encountered examples of work being done by sound involve the breaking of glass. Sonic booms, the sounds from explosions, and even the sound of a human voice can sometimes cause glass to break.

The phenomenon of hearing is another example of sound doing work. Sound waves generated by a human voice hit the eardrum, setting it in motion. The eardrum in turn vibrates three small bones in the middle ear that set the fluid in the inner ear in motion. The kinetic energy of this fluid is converted into electrical energy which is transported along the auditory nerve to the brain. These electrical impulses are interpreted by the brain as human speech.

Light energy is one form of **electromagnetic radiation**. Just as light comes in different colors, electromagnetic radiation comes in different forms. We call the part of the electromagnetic spectrum that our eyes are capable of sensing *light*; other parts of the same spectrum that we cannot see include radio waves, microwaves, ultraviolet rays (or black light), and X rays. All of these types of electromagnetic radiation are examples of the propagation of energy through space by waves in electric and magnetic fields. Unlike sound energy, electromagnetic radiation travels fastest and most efficiently through a vacuum.

Sunlight is the single most important energy source on Earth. All living things gain the energy necessary for life from sunlight. Every food chain begins with plants capturing the sun's energy. Plants use sunlight directly to power photosynthesis; animals harvest the sun's energy indirectly by eating plant carbohydrates and proteins. Even animals that are strictly meat-eaters are dependent on the sun for food: Meat-eaters survive by eating other animals that eat plants.

People also use the energy of sunlight that is stored in the Earth in the form of fossil fuels (coal, oil, and natural gas). These fuels provide energy for most of our industrial processes. Without sunlight, life as we know it could not exist on Earth.

Electrical energy can be considered either a form of kinetic energy or a form of potential energy. It is classified here as kinetic energy because electric current consists of charged particles moving through some type of conductive material. The amount of current moving along a wire (or other conductor) is measured in **amperes** (usually abbreviated A).

Most practical applications of electrical energy, such as lighting bulbs or running motors, involve the movement of electrons. Every electron has a charge of minus 1. A fundamental property of matter is that particles having the same charge repel each other, so electrons tend to move away from places that have high concentrations of negative charge and toward places that have low concentrations of negative charge.

This movement of electrons away from areas with a high concentration of negative charges to areas with less negative charge is analogous to the behavior of water flowing away from an overfilled lake to a low, empty streambed below.

However, the flow of electric current through the wires in your house is not exactly like the flow of the water in a creek. Electrons do not move in a straight line from a generator through the overhead utility to the wires and into your house. Utility companies distribute electrical energy in the form of alternating current, or AC, electricity. Alternating current generators change the direction of the flow of electrons 60 times per second. The electrons carrying energy to an electrical outlet in your home are actually oscillating back and forth along the wire.

Electrical energy is one of the most useful forms of energy because it is readily transportable by means of conductive materials (like copper or aluminum wire) and it can be used to perform an almost limitless array of tasks.

◆Gravitational, elastic, and chemical potential energy

Gravitational potential energy is the form of potential energy that is exploited in many of the activities in these modules. Any time a mass is raised in a gravity field, it gains gravitational potential energy. The equation for gravitational potential energy on Earth is stated:

GPE = weight x height.

The greater the height to which an object is lifted, the greater the amount of gravitational potential energy that will be stored in that object. If two objects are lifted to the same height, the heavier object of the pair will have more gravitational potential energy. (The demonstration in Module 3 comparing the impact of an apple falling 1 m with that of a brick falling 0.05 m shows the interrelationship of mass and height in gravitational potential energy.) The weight of an object depends on the strength of the gravitational force acting upon it. Therefore, the amount of gravitational potential energy stored by an object also depends on the strength of the gravitational force acting on that object. A ball being held 1 m above the surface of the moon will have only one-sixth the gravitational potential energy of an identical ball held 1 m above the Earth, because the moon's gravitational field is only one-sixth as strong as Earth's.

Pendula are excellent devices for demonstrating the conversion of gravitational potential energy to kinetic energy, and vice versa.

Before swing

Velocity = 0
Kinetic energy = 0
Gravitational potential energy is large

When a pendulum bob is being held above its rest position, it contains gravitational potential energy, but no kinetic energy.

During swing

Velocity increases
Kinetic energy increases
Gravitational potential energy decreases

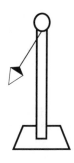

As it begins to swing downward, some of the potential energy is converted to kinetic energy; however, the bob still possesses both forms of energy.

Bottom of swing

Maximum velocity
Maximum kinetic energy
Minimum gravitational potential energy

At the low point of the swing, the bob will be moving at its greatest velocity, and essentially all of the original potential energy will have been converted to kinetic energy.

Rising swing

Velocity decreases
Kinetic energy decreases
Gravitational potential energy increases

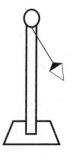

As the bob continues its swing and rises, more and more of the kinetic energy is reconverted to potential energy.

When it reaches the opposite end of its arc, the bob will stop moving for an instant. At that instant, it contains no kinetic energy, and all of its energy is stored in the form of gravitational potential energy. If there were no frictional forces affecting the pendulum, it would continue swinging back and forth forever.

Elastic potential energy is energy stored in the form of "springiness." All matter resists deformation—having its shape changed by a force. Brittle substances break rather than change shape if a strong enough force acts on them. "Springy" substances change shape and store at least some of the energy that was required to change their shape as elastic potential energy. When the force that deforms a springy substance is removed, its elastic potential energy is converted back into kinetic energy.

Toy balls, rubber bands, and metal springs all store elastic potential energy. Gases can also store energy in this form. You can sense the air in a bicycle pump resisting being compressed when you pump up a tire. If you block the exit tube on the pump, the pump handle will spring back after being pushed downward. The air's elastic potential energy is converted back into kinetic energy when the pump handle is released. The compressed air possesses the ability to do work—to move the pump handle through a distance.

Chemical potential energy is energy that is stored in the chemical bonds that hold atoms together to form molecules. In order to release chemical potential energy, chemical bonds must be broken.

The most common way to tap chemical potential energy is by burning a substance in the presence of oxygen. For example, when 4 g of hydrogen gas are burned, they will combine with 32 g of oxygen gas, producing a product of 36 g of water. This chemical reaction also produces 136,600 calories of heat.

Most of the energy required to maintain industrial societies comes from releasing the chemical energy stored in the bonds of some form of hydrocarbon compound (molecules containing both hydrogen and carbon). Coal, oil, natural gas, and wood are all hydrocarbon-containing substances that store large amounts of chemical energy. All of the energy stored in these hydrocarbons came originally from photosynthetic organisms that used the energy of sunlight to assemble the hydrocarbon molecules.

READING 7

Conserving Mass and Energy

◆Introduction

All the energy transformations that you have worked with in these modules obey two conservation principles: the *Principle of Conservation of Energy* and the *Principle of Conservation of Mass*.

These principles can be stated:
Energy can be changed from one form to another, but it cannot be created or destroyed.

Matter is not created or destroyed when a chemical reaction takes place.

◆The conservation of energy

In 1849, James Prescott Joule (for whom the unit of energy, the joule, is named) published a paper entitled *The Mechanical Equivalent of Heat*. This paper detailed a series of elegant experiments showing that heat and motion are different forms of a single phenomenon: kinetic energy. These experiments contributed data supporting the concept of the Principle of Conservation of Energy.

Scientists now take for granted the statement that energy is neither created nor destroyed. Energy "lost" during transformations can always be accounted for if you look for it "in the right place." Consider the case of a ball being held above a surface, then released. Some gravitational potential energy appears to be lost when the ball falls, since a ball never bounces all the way back to its starting point. However, finding the types of energy lost by a ball is a straightforward problem. On closer inspection we find that the lost energy is: transferred to the air (forming a tiny breeze); partially converted to sound (the thump you hear as it strikes the surface); and used to heat the ball, the air, and the surface that the ball strikes when it lands. The total amount of energy remains constant, but the form of this energy changes when the ball bounces.

If such a thing as a "perfect" energy-converting ball could be assembled, it would be able to convert all of its gravitational potential energy to kinetic energy and elastic potential energy, then reverse this process with 100% efficiency. This imaginary perfect ball would fall silently, release no thermal energy on impact, bounce back to exactly its starting height, and continue this cycle of bouncing forever. Making such a ball is impossible.

◆Conserving mass while undoing energy transformations

When 4 g of hydrogen gas and 32 g of oxygen are burned, they give off 136,600 calories of thermal energy. In accordance with the principle of conservation of mass, no new matter is created by this reaction, nor is any destroyed; the same atoms of hydrogen and oxygen are present before and after the reaction. The same amount of mass, 36 g, is also present before and after the reaction. Only the chemical bonds between the atoms have changed.

In principle, all types of energy transformations are reversible. In this case, if the equivalent of 136,600 calories of electrical energy is added to the chemical bonds of the 36 g of water, the bonds of the newly-formed water molecules will be broken. Four grams of hydrogen gas and 32 g of oxygen will be produced. These gases will be ready to burn again, and they can again release 136,600 calories of thermal energy. Mass and energy are conserved.

The process of breaking the bonds of the water molecules is not 100% efficient. More electrical energy will flow through the water than the 136,600 calories of chemical energy being restored to the hydrogen and oxygen. However, the electrical energy that appears to be lost can be accounted for by measuring the temperature of the water while electricity flows through it. The temperature of the water increases as the current flows; the lost electrical energy increases the thermal energy of the water.

Scientists have found that the principles of Conservation of Energy and Conservation of Mass hold true for all commonly-encountered energy transformations. Whenever energy is transformed, some of it is "wasted," but any energy that is wasted can be accounted for.

In the late 19th century, scientists began to investigate the behavior of atoms undergoing a previously unknown type of energy transformation. Their results indicated that these atoms were disobeying the time-tested conservation principles. Reading 8 examines this apparent exception to the rules of Conservation of Mass and Conservation of Energy.

READING 8

Nuclear Energy: An Exception to the Conservation of Mass and Energy?

◆Introduction

The principles of Conservation of Energy and Conservation of Mass form the foundation for classical studies of energy transformations. Until the late 1920s, most physicists considered these principles to be absolute. However, the conservation principles came to be questioned by physicists as they learned more about the structure of the atom. Early in this century, they discovered a new energy source that seemingly disobeyed the well established rules of conservation: **nuclear energy**.

◆Do the conservation principles apply when the nucleus changes?

The nucleus of any atom is composed of protons and neutrons. Early in this century, nuclear physicists making careful measurements of the masses of protons and neutrons discovered an unsettling fact: the nucleus of an atom possesses *less mass* than the individual protons and neutrons from which it is assembled. This discovery presented scientists with the difficult problem of having to explain why, when they looked closely at the particles composing the nucleus, they found that 1 + 1 equaled *less than* 2. What could explain the lost mass? Was mass being destroyed?

Similarly, scientists faced the problem of explaining the source of the energy released when heavy atoms break apart and form lighter elements (**nuclear fission**), or the nuclei of light elements are combined to form heavier elements (**nuclear fusion**). Enormous amounts of energy emanate from the nucleus of elements during these processes. Atom-for-atom, more than a million times as much energy can be released by nuclear reactions than by chemical reactions. At the turn of the century, there was no known source for this energy. Was energy being created?

In 1905, Albert Einstein published a simple equation that could explain these unsettling phenomena:

$$E = mc^2.$$

This equation means that energy is equal to mass times the speed of light squared. In other words, *matter and energy are interchangeable quantities*. Mass can be transformed into energy, and energy can be transformed into mass.

What does this tell us about the more conventional energy sources such as chemical potential energy, elastic potential energy, etc.? Does the nucleus operate under different rules with regard to mass and energy conservation?

On careful examination, the answer to this question turns out to be *no*. Even in everyday reactions like striking a match or storing elastic energy in a spring, mass and energy are interchangeable. A heated frying pan is slightly more massive than a cool one; a spring-driven cooking timer gains mass when it is wound.

In general, when energy is stored, mass increases; releasing energy decreases mass. However, the changes of mass in everyday energy conversions are so small that we do not notice them.

Nuclear energy obeys the same general principles as other energy forms; it releases large amounts of energy per atom per nuclear transformation because a significant "chunk" of the nucleus' mass is converted to energy.

The principles of Conservation of Energy and Conservation of Mass are still basically correct. However, they can be restated as the **Law of Conservation of Mass-Energy**, which accounts for the equivalence of mass and energy:

Neither matter nor energy can be created or destroyed, but they can be transformed into one another.

Einstein's insights fueled a revolution in physics that led scientists to view the universe in a totally new way. This new physics suggests different explanations for the structure and organization of our material universe. By revealing the energy of the nucleus, it has also provided a potential means to end all existence on Earth.

READING 9

Scientific Literacy for All

◆Introduction

We live in an age of scientific and technological innovation; every phase of our lives is touched by the products and processes of these two enterprises. While pre-college preparation of future scientists is vitally important, training future scientists cannot be the sole function of a K–12 science program. Our technological society demands that *all* future citizens be scientifically literate.

As our increasingly sophisticated technology provides us with new industrial capabilities and more personal leisure time, it also confronts us with new sets of problems. Important decisions must be made about implementing new technologies, funding basic research, and allocating Earth's resources. Citizens *must* have a working knowledge of science and mathematics in order to participate intelligently and responsibly in the decision-making process.

Several national groups and numerous individuals have attempted to define scientific literacy. These definitions are useful because they give direction to those who teach and develop instructional materials. Several of these definitions have analyzed scientific literacy in terms of the behaviors desired of a scientifically literate person. For example, according to one definition,* the scientifically literate person:

• has knowledge of the major concepts, principles, laws, and theories of science and applies these in appropriate ways.

• uses the processes of science in solving problems, making decisions, and other suitable ways.

• understands the nature of science and the scientific enterprise.

• understands the partnership of science and technology and its interaction with society.

• has developed science-related skills that enable him or her to function effectively in careers, leisure activities, and other roles.

• possesses attitudes and values that are in harmony with those of science and a free society.

• has developed interests that will lead to a richer and more satisfying life, and a life that will include science and life-long learning.

There are many ways to develop each of the above components of scientific literacy while teaching mechanics. A given activity, reading passage, or film often contributes to several components. It is not necessary to have students do one activity to teach them certain principles of science, another activity to teach processes of science, and another to show how science affects our society.

The activities in these modules use the processes of science to teach problem solving and to encourage the development of critical thinking skills. For example, building a "tardiness preventer" for Activity 19 encourages students to work together as a research and development team. In order to succeed, they must know the basic principles of simple machines, design and build an original complex machine, observe how their machine works (or doesn't work), and modify the machine on the basis of their observations and inferences.

As you use these activities with your students, you can also provide information about the concepts being studied through written materials,

*Simpson, R. D. and Anderson, N. D. (1981). *Science, Students, and Schools: A Guide for the Middle and Secondary School Teacher.* New York: John Wiley and Sons.

films, lectures, and discussions. Together these reveal the nature of science: a dynamic partnership between knowledge and process.

◆Equity in education

Equity in educational, employment, and housing opportunities is a familiar topic to most Americans, largely because of the prominence of civil rights issues in the past two decades. Formerly, there were very few opportunities for women and minorities to have careers in science and science-related fields such as engineering and medicine. This is no longer true. However, these groups still encounter many barriers to full participation in science, often because of sex- and race-role stereotyping.

Researchers have found that when boys and girls are working together in groups, the boys will set up the apparatus and do the experiment while the girls take notes and record the results. In addition, teachers often ask boys harder questions than they ask girls, give boys more time to think about their answers, and in other ways communicate that they have higher expectations for boys than for girls. Instructional materials frequently ignore the contributions of women and minorities in science, or portray these groups in stereotyped ways, thus depriving female and minority students of appropriate role models.

Equity in science education means encouraging female and minority students, as well as white males, to participate actively in science lessons. Teaching styles that are supportive of an equal opportunity classroom become especially important in light of the changing demographics of the American populace. The composition of U.S. citizenry is shifting. The birth rate of the American white population is declining, whereas that of the non-white population (particularly Hispanic Americans) is increasing rapidly. "Minorities" are rapidly becoming the majority of the student population in many places. It is, therefore, even more important that all American youth receive the very best education available.

Educational research has begun to identify some of the components of effective and equitable teaching. For example, during the past decade researchers have learned that there are many beneficial outcomes when students work in cooperative groups on a task with each person contributing to the learning of the others. Most importantly, noncompetitive methods seem to make learning more accessible to a wide range of students. Advanced students develop their leadership skills while they reinforce their subject knowledge. Students who learn at a less rapid rate have time to discover answers via trial and error, and to observe the logical processes demonstrated by their peers. In general, cooperative learning activities encourage students to develop learning skills by providing a supportive climate where students can learn with less risk of embarrassment or failure.

No one advocates that all lessons be conducted in this way; there is a place for individual work, for drill and practice, for whole-group discussions, and for other teaching and learning methods. However, cooperative group learning is particularly well-suited for use in laboratory settings and for carrying out some of the other activities described in these materials.

Students who have spent their school years in classrooms where cooperation was not encouraged or even allowed will not be able to work cooperatively without training in this kind of group work. It takes time and practice to learn to work interdependently. If you are interested in learning more about the positive outcomes produced by this kind of teaching, you may wish to read *Circles of Learning: Cooperation in the Classroom*, by D. W. Johnson and R. T. Johnson, (Alexandria, VA: Association for Supervision and Curriculum Development, 1984); *Gifted Young in Science: Potential Through Performance*, edited by P. F.

Brandwein, A. H. Passow, et al., (Washington, DC: NSTA, 1989); and *What Research Says to the Science Teacher, Volume 5: Problem Solving*, edited by Dorothy L. Gabel, (Washington, DC: NSTA, 1989).

READING 10

Opening Up the Black Box of a Complex Machine

◆Introduction

Schools and homes are filled with both complex and simple machines. All too often we simply take these machines for granted, or treat them as black boxes—devices with some incomprehensible inner mechanism that functions in some mysterious way. We don't understand how these machines work; we just expect them to do what we want.

However, with a little knowledge about simple machines and an understanding of the basic principles of Newtonian mechanics, you can figure out how even complicated machinery works.

◆Automatic seat belts: Opening a black box

Most automobiles built since 1970 have some version of an automatic seatbelt. These seatbelts hold you firmly in place *only* during sudden braking or during a collision. The mechanism of such belts is an example of a black box. You probably use seatbelts like this, but do you know how they work?

This type of belt is built from simple machines. Many automatic seatbelts have only three main parts: a pulley, a lever, and a pendulum that activates a locking mechanism. The belts grab you in an emergency because the pendulum obeys Newton's first law of motion: An object at rest tends to stay at rest, and an object in motion tends to stay in motion in a straight line at a constant speed unless acted upon by an unequal force.

The following article, reprinted from the March, 1985 JETS Report, explains how this works. The last paragraph in the article notes that additional states were considering adopting laws that require motorists and passengers to wear seatbelts. As of March 5, 1990, 33 states and the District of Columbia have mandatory seatbelt laws in effect for adults. All states and DC have mandatory seatbelt laws for children.

How Do Seat Belts Work?*

The front seat belts in newer cars unwind easily and seem to have very little tension. So how can they hold you securely during an accident? They do it with the help of a mechanism based on a simple fact of physics.

Seat belts, or shoulder belts, are designed to restrict your movement only in an emergency. During normal driving conditions. you can lean forward or to the side with very little pressure from the belt.

The mechanism that makes all this possible is called an *inertial reel*, a device as simple and reliable as gravity. An inertial reel has three major moving parts: a ratchet, a tilting lever that can lock the ratchet, and a pendulum that nudges the lever into locking position. Each seat belt rolls up on a spring-wound spool connected to the ratchet.

Now let's look at that principle of physics we mentioned. It's *inertia*, the tendency of a body at rest to remain at rest, or of a body in motion to stay in motion in a straight line unless disturbed by an external force.

As long as the car is moving normally or standing still, the pendulum hangs vertically and the spool can unwind. But in a sudden stop, the lower, heavy end of the pendulum swings forward; because of inertia, it "wants" to keep moving. The upper end of the pendulum then strikes a

*Reprinted from the March, 1985 JETS Report by permission of the Junior Engineering Technical Society, Alexandria, Virginia.

lever, the locking bar, which locks the ratchet and secures the belt.

Dreadful Data . . . 1,501 people were killed in traffic crashes in one state during 1984. Of those killed, 928 had safety belts available in their motor vehicles, yet 863 chose not to buckle up. That means the percentage of non-use is 93%. Other states report similar percentages.

How the inertial wheel works. *Your shoulder belt is designed to allow freedom under normal conditions, but to lock automatically and restrain you in a collision.*

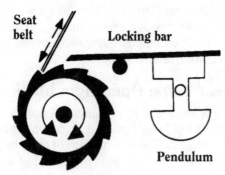

Seat belt

Locking bar

Pendulum

Ratchet mechanism

Under normal conditions, *the pendulum and locking bar are in their rest positions. The reel which holds the seat belt is free to rotate. As you lean against it, the belt unreels.*

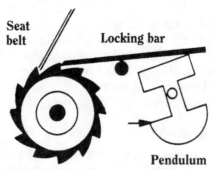

Seat belt

Locking bar

Pendulum

Ratchet mechanism

In emergencies, *such as a collision from any direction, the pendulum tilts, forcing the locking bar to engage the ratchet. The reel locks and the seat belt restrains you.*

MATERIALS AND SOURCES

Guide for Teachers and Workshop Leaders

This section contains module-by-module master lists of materials, equipment, and audiovisual materials (if any) recommended for use in each activity, and gives ordering information about the audiovisual materials recommended for use with this manual. Substitutions for some materials may be suggested in the "Preparation" section of each activity.

Master Lists of Materials

◆Module 1:
Pondering Projectile Pathways

Master list of materials and equipment for each group:

fifteen to twenty 3 x 5 inch index cards

115-cm length of clear plastic tubing (1" OD, 3/4" ID)

a "bullet"—a marble or ball bearing that can roll freely inside the tubing

a meter stick

a sheet of paper measuring at least 60 cm x 60 cm

a felt tip pen or other marker

a tennis ball

2 popsicle sticks

a plastic cereal bowl or a half sheet of paper

goggles or safety glasses with side shields for each member of the group

a piece of 2" x 4" lumber about 15 cm long

a medium-wide to wide rubber band at least 4 mm wide and 8.5 cm long

clay

a large hard-back book

a protractor

The following may be used by the entire class or workshop group:

masking tape

a ball of string

◆Module 2:
It's All in a Day's Work

Master list of materials and equipment for each group:

a common brick (preferably one with holes through it)

a smooth surface to drag the brick across—a lab table or tile or vinyl floor

2 meter sticks

a spring scale calibrated in newtons or grams

3 ice cubes

safety glasses or goggles (for each member of the group)

a Handi-Mates™ replacement tissue roller #02511 by Mirra-Cote™ (or a similar plastic roller from another manufacturer)

a fine-toothed handsaw (a coping saw or hacksaw will work well)

a small piece of medium-grit sandpaper

a drill with 1/8" bit

a fine-point permanent black marking pen

a large marble (a "shooter" approximately 2.5 cm in diameter)

a section of N-gauge model railroad track approximately 75 cm long

several large books

a protractor

Eberhard Faber Holdit™ plastic adhesive—2 grape-size balls

a laboratory balance calibrated in grams

The following may be used by the entire class or workshop group

a ball of string

masking tape or duct tape

optional: a paper clip and candle (instead of the drill)

Audiovisual Equipment:

VHS videocassette player and television

Eureka! video series program #8

◆Module 3:
Energy and Energy Conversions

Master list of materials and equipment for each group:

2 ball bearings (1/2" in diameter)

plastic tubing (at least 1 m in length; 3/4" inside diameter)

a rubber band (about 0.6 cm x 9 cm)

a standard-size wire paper clip

a brick

an apple of average size

a meter stick

a small, empty soft drink bottle (16 oz or smaller)

a balloon

baking soda (sodium bicarbonate)

100 ml of vinegar

a spring-driven "jump up" toy (or a "wind up" toy)

a Superball™ or similar high-bouncing ball

a squash ball (a Slazenger™ standard ball is recommended)

a thermometer (a laboratory thermometer calibrated to 100°C)

tongs

optional: a small funnel

The following can be used by the entire class or workshop group:

paper towels

a container of ice water (mostly ice with a little liquid)

a container of hot water (temperature should be about 50°–60° C)

Audiovisual Equipment:

VHS videocassette player and television

Eureka! video series programs #9 and #10

◆ Module 4:
Machines Are Simple!

Master list of materials and equipment for each group:

a toy car, toy truck, or similar object that rolls and to which a spring scale can be attached. The car should weigh 3–4 N.

a smooth board that is wide enough for the car to roll on. It should have a minimum length of 122 cm.

a spring scale (calibrated to 5 N)

a meter stick

several large books

5 lead weights (2-oz fishing sinkers)

1 standard wire paper clip

1 small binder clip

an uncooked egg

a pan of water

The following may be used by the entire class or workshop group:

masking tape

a ball of heavy twine

sewing thread

Audiovisual Equipment:

VHS videocassette player and television

Eureka! video series programs #11, #12, and #13

◆ Module 5:
Seesaws, Steering Wheels, and Screwdrivers:
Applications of Torque

Master list of materials and equipment for each group:

two cups of water, both filled to the top

a 2" x 6" plank 3 m long (this is the preferred size; the minimum size acceptable is a 2" x 4" that is 2.5 m long)

access to a door

2 common bricks (with holes in them)

a large dowel (it must fit through the holes in the bricks; the length should be approximately 0.75 m to 1.25 m)

a broom

a paper cup or other small, unbreakable object

a yo-yo

several average-size screws

a board with 6 or more holes drilled into it (the screws should fit the holes tightly)

2 identical, average-size screwdrivers

2 small blocks of wood (about 2 cm x 2 cm x 5 cm)

a hammer

a 35-cm length of 2" x 4" lumber

2 nails (6d or larger common nails)

2 heavy-duty rubber bands

a medium or medium-small rubber band

a metric ruler or meter stick

optional: a 5-cm length of masking tape

The following may be used by the entire class or workshop group:

duct tape

masking tape

◆ Module 6:
Center of Gravity

Master list of materials and equipment for each group:

access to a wall

2 chairs (at least one of which must have a back)

a lead weight (a 1-oz fishing sinker)

a straight edge

a sharp pencil

enough photocopies of the Cynthia pattern (see page 131) for all groups

a cylinder (a large holiday cookie container)

a large weight (several 2-oz fishing sinkers)

an inclined plane (a board propped up on books)

40 cm of fairly stiff wire or a wire coat hanger

a lump of clay about the size of a golf ball

an empty cereal box

2 styrofoam cups and lids

1 cup of salt (or dry, fine sand)

enough photocopies of the Motorcycle Michael and Balancing Belinda patterns (see pages 139 and 140) for all groups

scissors

a stapler

3 average-size plastic drinking straws with flexible elbows

1 rectangular piece of lightweight cardboard measuring 4 cm x 28 cm. (The piece can be cut from the cardboard back of a tablet or from posterboard.)

a metric ruler

12 (or more) paper clips of the same size

The following may be used by the entire class or workshop group:

a ball of sewing thread or very light string

masking tape

a ball of string

Recommended Audiovisual Materials

Depending on the needs, interests, and abilities of your students, these audiovisual materials may provide useful supplements or alternatives to textbook or lecture presentations of selected concepts of Newtonian mechanics.

◆*Eureka!* video series

Eureka! is an animated video series produced by TVOntario (© 1981). It uses examples drawn from everyday experiences to demonstrate the behavior of matter in motion. The cartoons are useful for introducing or reviewing the concepts of mechanics.

Many school districts already have copies of *Eureka!* available for use. Check with your district or state instructional TV personnel to see if yours is one of them.

For information on how to rent or purchase copies of *Eureka!* contact:

TVOntario
Suite 206
143 West Franklin Street
Chapel Hill, North Carolina 27514
(800) 331–9566 or (919) 967–8004 (in NC)

Metric Conversions

Only two countries in the world, Burma and the United States, persist in using the English system of measurement.* Although some U.S. industries are beginning to build items to metric standard measurements in order to become more competitive in international markets, the old units of feet, pounds, quarts, and other English measures are still commonly used for many consumer items in the United States. Until we join England in giving up the English system, converting from English to metric units or vice versa will sometimes be necessary. The following table tells how to do the most common conversions:

Rules for English to Metric Conversions

Multiply	by	this number		to get
_____inches	X	2.540	=	_____centimeters
_____feet	X	0.305	=	_____meters
_____miles	X	1.609	=	_____kilometers
_____quarts	X	0.946	=	_____liters
_____gallon	X	3.784	=	_____liters
_____pounds	X	0.454	=	_____kilograms
_____°F − 32°	X	0.556	=	_____°C

*Popular Science, Vol. 233, No. 1 (July 1988, p. 45, "What's News").

MECHANICS II

Glossary

Acceleration: The rate at which an object speeds up or slows down. By Newton's second law, in order for an object to be accelerating, a force must be acting on it. Acceleration is occurring if an object travels *different distances* during equal time intervals that its motion is observed. The units for acceleration are meters per second per second, or meters per second squared (m/s^2).

Air resistance: A force exerted on a moving object opposite to its direction of motion due to the friction between the object and air. Air resistance is also called *drag* or *air friction*.

Center of gravity: The point through which the force of gravity for an entire object may be considered to act, regardless of the object's orientation. In other words, the center of gravity is the point where we may consider all of an object's weight to be located.

Chemical potential energy: Potential energy that is stored in the chemical bonds that hold an object's molecules together.

Dynamics: A branch of mechanics that concentrates on the study of bodies in motion.

Elastic potential energy: Potential energy that is stored in the molecules of objects that are stretched or compressed.

Energy: The ability to do work. Something contains energy if it can apply a force through a distance.

Force: A push or pull in a particular direction that can be applied to an object. We can more technically define a force as something that has the capacity to change the motion of an object. Both the *magnitude* (strength) and the *direction* must be stated when defining a force. Vectors are often used to represent forces.

Friction: Resistance to relative motion between objects in contact. The force due to friction acts on an object in the direction opposite to that of its motion.

Gravitational potential energy: Potential energy that is due to an object's position in a gravity field.

Gravity: A phenomenon that exists throughout the universe. The force due to gravity is the force of attraction that exists between all objects in the universe. The amount of gravitational force between two bodies (such as the Earth and a rock thrown up into the air) depends on the mass of both objects and the distance between them. Earth's gravitational force is just one example of the general phenomenon of gravity. On Earth, the force due to gravity is the force that causes objects (such as an airborne rock) to accelerate toward the Earth.

Inertia: A measure of an object's resistance to change in motion. Inertia is another way of describing an object's mass. The more mass that an object possesses, the more force that is required to set it in motion or to stop it from moving. Inertia is a property possessed by all matter that can be thought of as laziness or "difficult-to-moveness."

Joule: The preferred unit of measurement for work in the metric system. One joule (1 J) is the amount of work done when a force of 1 N is applied through a distance of 1 m.

Kinetic energy: The energy of moving objects.

Machine: An object that changes the force necessary to perform some task.

Mass: A property of matter related to inertia. As the mass of an object increases, so does its inertia. Mass can be thought of as the quantity of matter in an object. Mass is *not* the same as weight.

Mechanical advantage: A measure of the advantage gained by using a machine to change the force necessary to perform a task. Mechanical advantage is calculated by dividing the force that the machine produces by the force applied to the machine.

Mechanics: The branch of physical science that describes the behavior of bodies in motion; mechanics deals with energy and forces and their effects on bodies.

Newton: The standard metric unit of force. One newton (1 N) is the amount of force required to accelerate a 1000-g mass at a rate of approximately 1 m/s².

Potential energy: The energy stored in an object due to its position or the arrangement of its parts.

Projectile: An object cast or thrown by an external force. Its motion continues because of its own inertia.

Speed: The rate of motion; speed combines information about how far an object travels with how long it takes to travel that distance. An object's speed is calculated by dividing the distance traveled by the time interval. The metric units for speed are meters per second (m/s). Speed only tells us about the magnitude of motion, not the direction. An object moving at a *constant speed* will move the same distance during each successive time interval.

Torque: A quantity that is a measure of the turning effect of an applied force; it is the product of the applied force and the perpendicular distance from the point of application to the axis or point of rotation of the object being turned. A torque is applied any time an object is turned or rotated.

Vector: An arrow that can be used to represent quantities such as force or velocity. The *head* of a vector shows direction; the *tail* shows the magnitude or strength. A long tail indicates a strong force; a short tail indicates a weak force.

Velocity: A measure of motion that specifies both the *direction* of motion and the *magnitude* of motion. Velocity is *not* the same as speed, because the speed of an object tells how fast the object is moving, but does not specify the direction in which it is moving. Speed is actually the magnitude of velocity.

Weight: On Earth, a downward force that acts on an object due to the gravitational attraction between the object and Earth. *Weight is not the same as mass.* On the moon, you would only weigh 1/6 as much as you do on Earth, because the moon's gravitational attraction is less than Earth's. You would still have the same amount of mass, however, even though you would weigh less on a spring scale.

Work: The product of an applied force and the distance through which it acts.